STRANGE MAPS

STRANGE ƧꓯM

AN ATLAS OF **CARTOGRAPHIC** CURIOSITIES

FRANK JACOBS

VIKING STUDIO

VIKING STUDIO
Published by the Penguin Group
Penguin Group (USA) Inc., 375 Hudson Street, New York, New York 10014, U.S.A.
Penguin Group (Canada), 90 Eglinton Avenue East, Suite 700, Toronto, Ontario,
Canada M4P 2Y3 (a division of Pearson Penguin Canada Inc.)
Penguin Books Ltd, 80 Strand, London WC2R 0RL, England
Penguin Ireland, 25 St. Stephen's Green, Dublin 2, Ireland
(a division of Penguin Books Ltd)
Penguin Books Australia Ltd, 250 Camberwell Road, Camberwell,
Victoria 3124, Australia (a division of Pearson Australia Group Pty Ltd)
Penguin Books India Pvt Ltd, 11 Community Centre,
Panchsheel Park, New Delhi – 110 017, India
Penguin Group (NZ), 67 Apollo Drive, Rosedale, North Shore 0632,
New Zealand (a division of Pearson New Zealand Ltd)
Penguin Books (South Africa) (Pty) Ltd, 24 Sturdee Avenue,
Rosebank, Johannesburg 2196, South Africa

Penguin Books Ltd, Registered Offices:
80 Strand, London WC2R 0RL, England

First published in 2009 by Viking Studio,
a member of Penguin Group (USA) Inc.

3   5   7   9   10   8   6   4

Portions of this book first appeared in the Strange Maps blog.

Illustration credits appear on pages 235-240.

ISBN 978-0-14-200525-5

Printed in Singapore
Set in Sabon MT and Akzidenz Grotesk Pro
Designed by Renato Stanisic

While the author has made every effort to provide accurate telephone
numbers and Internet addresses at the time of publication, neither the publisher
nor the author assumes any responsibility for errors, or for changes that occur after
publication. Further, publisher does not have any control over and does not assume
any responsibility for author or third-party Web sites or their content.

FOR HANNE—THE LEGEND ON MY MAP

CONTENTS

INTRODUCTION
NOT FOR NAVIGATION

A word of warning: This is the most improbable, incomplete and incorrect atlas you're ever likely to hold in your hands. But it's also—hopefully—one of the funniest, most surprising collections of maps ever to be contained within the covers of a book . . . and somewhat instructive to boot.

This book grew out of a love for maps, a love mirrored by a dislike for atlases—*regular* atlases, that is. Every regular atlas presents a variation on the same themes: physical and political maps of the world, its continents and countries, maybe with some astronomy or socioeconomic geography thrown in to distinguish it from other atlases. For in essence, all regular atlases tell the same old story. Imagine going to a bookstore or the library to pick up a riveting read, and all they have is thousands of copies of *The Da Vinci Code*. Things would get boring pretty soon.

That's why I love maps, *strange* maps, the kind you won't find in a regular atlas. I love the unfamiliar, fascinating stories they tell. And that's why in September 2006 I started a weblog I called Strange Maps. Its mission: to collect cartographic curiosa that drifted around on the Internet, and to boldly go where no atlas had gone before.

Never did I imagine that Strange Maps would be popular beyond a small circle of fellow map geeks. But I was wrong. When I was a kid, I was the only one I knew who read atlases for fun (before I got bored with them); exploring this niche online has revealed that I'm far from alone. By March 2009, the blog had garnered over 10 million hits, which makes it one of the biggies in the blogosphere. It has been discussed on hundreds of Web sites and other blogs, and was featured in print by "old" media from Slovenia to the United States, from Britain to Brazil. It seems that if, to paraphrase Emerson, one manages to find an interesting niche, the world will make a beaten path to your blog.

The blog's success revived what had been a dormant idea from the start, an idea that seemed quite improbable at the time: to produce a real-world, hard-copy version of the blog—an actual Atlas of Strange Maps, an *anti-atlas* of sorts (geography buffs might appreciate the double entendre). I'm grateful to the good people at Viking Studio Press in New York for the opportunity to do so.

But this book remains an anti-atlas on several levels: Not only does it include maps excluded by more mainstream atlases, it also cannot aspire to those volumes' claim to completeness. The scope of curious cartography is bounded only by the imagination of cartographers, professional and amateur, everywhere. You might have a strange map in mind and vainly look for it in this book. But the ones that made it into this volume will I hope surprise and delight you nonetheless.

Another crucial difference: This atlas is not for navigation! These maps were not selected for their verisimilitude or fiability, but for their strangeness. Which brings me to a question that I've long pondered, without arriving at a watertight definition: What exactly *is* a strange map?

Selecting these maps has been a subjective,

eclectic and personal endeavor. This selection includes some maps that first appeared in . . . other atlases. Isn't that breaking the number one criterion of an anti-atlas? Well, no. Those maps might have featured in ancient atlases, but wouldn't be included in modern ones.

Criterion number one for a map to be strange is that something about them should strike me as being out of the ordinary. It might be the rarity of their subject matter: cartographic misconceptions, ephemeral states forgotten by history, bizarre idiosyncrasies of certain boundaries. The striking element might also be the treatment: the wild cartographic exaggerations of parody and propaganda, the skewed perspective of artists, maps forming the necessary background for literary invention. Or it might be the method: statistics masquerading as maps, revealing in one glance some information about economy, culture, linguistics and so on. It might be anything, as long as it is . . . strange. How's that for circular logic?

The maps in this book have been herded into categories, and although this juxtaposition yields some interesting insights, the subdivision is, again, rather subjective. But beyond all its eclecticism, subjectivity and haphazardousness, this collection also has a more serious raison d'être. Every strange map tells a story—sometimes surprising, always informative and generally not covered by those more comprehensive atlases.

Isn't it time you learned about Amikejo, the world's first and only Esperanto state? Wouldn't it be nice to know where to head for if you want to see a total solar eclipse between now and 2020? Haven't you ever wondered what Europe would have looked like if the Nazis had won?

If not, then there are over a hundred other maps in this atlas, waiting for you to pore over them, to read them like a story, to be surprised, shocked or amused by them. Enjoy!

Frank Jacobs
Hasselt/London, October 2008
Find the Strange Maps blog at:
http://strangemaps.wordpress.com

I. CARTOGRAPHIC MISCONCEPTIONS

In cartography, precision is essential.
But imagination can be an entertaining substitute.

Table of the COMPARATIVE HEIGHTS of the PRINCIPAL MOUNTAINS &c. in the World.

REFERENCE.

N. AMERICA.		Feet.	S. AMERICA.		Feet.	EUROPE.		Feet.	ASIA.		Feet.	AFRICA.		Feet.
1 Popocatepetl	Mexico	17.710	22 Chimboraço Highest of the Andes		21.441	39 Mont Blanc Highest of the Alps		15.665	60 Dhawalgeri the highest of the Himmaleh			76 Atlas Mtns	Morocco	12.500
2 Orizaba		17.371	23 Disca Cassada		19.570	40 Rosa		15.540	Mtns & the highest in the World. Hind.		26.462	77 Peak of Teneriffe		12.358
3 Mt St Elias	N.W. Coast	12.680	24 Antisana		19.249	41 Cervin		14.780	61 Yamaturi a Peak of the Himmaleh Mtns		25.500	78 Nieuveldt Mtns	S. Africa	10.000
4 Long's Peak	Rocky Mtns	12.500	25 Catopaxi		18.891	42 Schreckhorn		13.000	62 Dhaibun		24.740	79 Gross Morne	I. of Bourbon	9.000
5 James's do.		12.000	26 El Altar		17.256	43 Glockner	Germany.	12.130	63 Inferior summits of the Himmaleh			80 Mt of Gondar	Abyssinia	8.450
6 Volcano de Colima	Mexico	9.186	27 Minissa		17.238	44 Perdu Highest of the Pyrenees		11.265	Mtns varying from 24.500 to		19.000	81 Mt Taranta		7.800
7 City of Toluca		8.818	28 Sangai		17.136	45 St Bernard		11.000	64 Mt Ophir	I. of Sumatra	13.800	82 Schneeberg Mtns	S. Africa	6.390
8 Mexico		7.470	29 Tunguragua		16.500	46 Simplon		11.000	65 Mt Ararat	Armenia	12.000	83 Kamberg		5.644
9 Durango		6.847	30 Pichinca		15.939	47 Etna	Sicily	10.700	66 Altai Mtns highest peak	Russia	10.700	84 Table Mt		3.831
10 White Mtns Highest Peak	N. Hamp.	6.634	31 El Corazon		14.790	48 St Gothard	Switzerland	9.500	67 Aratscha	Komtschatka	9.600	85 Diana's Peak	I. of St Helena	2.700
11 Moosehillock		4.636	32 Farm House of Antisana the highest			49 Lommin	Hungary	8.640	68 Lebanon	Palestine	9.535	86 The principal Pyramid	Egypt	500
12 Mansfield Mt	Ver.	4.279	inhabited spot in the World		13.434	50 Velino Highest of the Appenines		8.387	69 Hermon		8.949			
13 Camels Hump		4.188	33 Plain of Assuay		13.123	51 Olympus	Greece	6.500	70 Gete	I. of Java	8.500			
14 Saddle Back	Mass.	4.000	34 Boueran		12.652	52 Hecla	Iceland	5.000	71 Peak of Quaelpert		6.400			
15 Table Mt	S. Car.	4.000	35 Mines of Chota		11.562	53 Ben Nevis	Scotland	4.387	72 Ural Mtns highest peak	Russia	4.900	Reference to the Colours		
16 Peaks of Otter	Vir.	3.955	36 City of Quito		9.514	54 Ben Lawers		4.015	73 Ghauts	Hindoostan	4.000	The Mtns of N. America are coloured		Blue
17 Round Top	N. York	3.804	37 Santa Fe de Bogota		8.964	55 Vesuvius	Naples	3.731	74 Mt Tabor	Palestine	3.000	S. America		Yellow
18 High Peak		3.718	38 Popayan		5.905	56 Snowdon	Wales	3.571	75 Mt Carmel		2.000	Europe		Red
19 Grand Monadnock	N. Hamp.	3.254				57 Macgillicuddys Reeks	Ireland	3.404				Asia		Green
20 Alleghany Mtns average height		2400				58 Crossfell	England	3.390				Africa		Brown
21 Blue Mtns	Conn.	1.000				59 Skiddaw		3.175						

1

Nice Colors, Wrong Height: Mapping Mountains in 1831

This map is hopelessly outdated, but deliciously colored. It shows the world's tallest mountains—at least as they were known in the year 1831—arranged and color-coded per continent, and listed by height at the bottom.

The North American mountains are tinted blue (and numbered 1–21). The Mexican Popocatépetl is listed as the highest in North America, at 17,710 feet (5,398 m). In fact, we now know that Mount McKinley (also known as Mount Denali) in Alaska is the record holder, at 20,320 feet (6,194 m). Popocatépetl's height has been corrected to 17,802 feet (5,426 m), making it lower than the corrected figure for the Pico de Orizaba (18,490 feet; 5,636 m), still second on this list.

Ecuador's tallest mountain, Chimborazo, was considered the world's tallest until the beginning of the nineteenth century. It has been duly downgraded on this map, but still holds the distinction of being the tallest of South American mountains. That distinction should go to Mount Aconcagua in Argentina (22,841 feet; 6,962 m). To add insult to injury, Chimborazo's height has been downgraded from 21,441 to 20,565 feet (6,535 to 6,268 m).

Europe's mountains are tinted red. Mount Elbrus in the Russian Caucasus is nowadays considered Europe's highest (at 18,510 feet; 5,642 m), but as the geographic definition of Europe was a bit narrower at the time of this map, the honor goes to Mont Blanc (15,665 feet or 4,775 m, since adjusted to 15,781 feet or 4,810 m).

The tallest of Asia's mountains (colored green) is named as Dhawalgeri. Usually spelled Dhaulagiri today, this Nepalese mountain was thought of as the world's highest for thirty years after its discovery in 1808, and actually is the world's seventh tallest. Its height has been adjusted from 26,462 to 26,795 feet (8,066 to 8,167 m). Mount Everest's highest mountain status wouldn't be established until 1852 (it would be named in 1865).

African mountains are colored brown. Relatively mountainless, Africa's highest peak is Mount Kilimanjaro (at 19,340 feet or 5,895 m also the world's tallest freestanding mountain). Africa's highest point is misidentified on this map as Mount Atlas in Morocco (12,500 feet; 3,810 m).

A few other interesting features of this map:

- It shows ten active volcanoes, spewing fire and smoke.
- Charmingly, at number 32, the map also indicates the world's highest inhabited spot: a farm house at Antisana, in South America.
- For comparison's sake, number 86 is Giza's principal pyramid (its height given as 500 feet [152 m]; actually, the Great Pyramid was only 480 feet (146 m) when it was still complete).

2

Black Rock, Fake Rock: The First, False Map of the True North

Sometime in the fourteenth century, a Franciscan from Oxford, a "priest with an astrolabe," wrote a travelogue about his discoveries in the North Atlantic, the *Inventio Fortunata* (The Discovery of Fortunata or The Fortunate Discovery) and in 1360 presented it to the king of England. This book has been lost since the late fifteenth century.

However, a Jacobus Cnoyen from the city of 's Hertogenbosch (in the present-day Netherlands) summarized the contents of the *Inventio*, related to him in 1364 in Norway by another Franciscan who had met the work's author. Cnoyen's own travel book was called the *Itinerarium*. This has also been lost.

All this we know by the extensive quotes from the *Itinerarium* in a letter by the Flemish cartographer Gerardus Mercator to his friend the English scientist, occultist and royal adviser John Dee. That letter, written in 1577 and now in the British Museum, mentions that "in the midst of the four countries is a Whirl-pool, into which there empty these four indrawing Seas which divide the North.

And the water rushes round and descends into the Earth just as if one were pouring it through a filter funnel. It is four degrees wide on every side of the Pole, that is to say eight degrees altogether. Except that right under the Pole there lies a bare Rock in the midst of the Sea. Its circumference is almost 33 French miles, and it is all of magnetic Stone. . . . This is word for word everything that I copied out of this author [i.e., Cnoyen] years ago."

A giant magnetic rock, exactly at the North Pole . . . well, that *would* explain why all compasses point north, wouldn't it? Alas, the ominous magnet (described in the letter as "black and glistening" and "high as the clouds") is a bit too fantastic an explanation for the phenomenon of magnetism. For even back in the late sixteenth century, mariners often found that their compasses increasingly deviated from "true north" as they approached it.

But only later did the separate (and wandering) location of the magnetic poles become common knowledge. In the intervening Age of Exploration (and sometimes Fabulation),

Mercator cited an author who clearly hadn't seen the North Pole with his own eyes—nor had the author *he* quoted, nor in fact would anyone for centuries to come.

In the meantime, the invented geography in the *Inventio Fortunata* that came to us via that one letter greatly influenced cartographers' views of the Arctic region. For if no other knowledge of yet undiscovered lands is available, there's really not much argument against unbelievable stories.

And so the Black Cliff, the four countries and the whirlpool are evident in Martin Behaim's globe (1492), which predates Mercator's map. In 1956, a letter surfaced written by the English merchant John Day in 1497 or 1498 to "the Lord Grand Admiral" (probably Columbus). In it, Day expressed regret that he hadn't been able to find the *Inventio Fortunata* for him. In a marginal note on one of Johannes Ruysch's maps (from 1508), the Dutch cartographer even mentions that two of the continents surrounding the North Pole are inhabited.

Mercator's late-sixteenth-century Arctic

map Septentrionalium Terrarum (Of the Northern Lands) was the first ever to be centered on the North Pole itself. It was a mix of fact and fiction, showing some recent discoveries but also the four fanciful countries surrounding the Arctic whirlpool within its middle the *Rupes Nigra et Altissima* ("Black and Very High Cliff"), supposedly responsible for animating navigators' compasses.

On the subject of mixing fact with fiction, Mercator incongruously includes in his map two *other* magnetic poles, along the 180 degrees meridian, indicating that he *did* know of the magnetic deviation from the "true north," but wasn't yet prepared to ditch the preceding fabulation.

Mercator's map was included in the last of three volumes constituting his groundbreaking work (the first geographic tome to be called an atlas). The cartographer didn't live to see it published: The last volume was brought out by his son Rumold in 1595, the year after his death.

In 1604, the cartographer Jodocus Hon-

dius acquired the printing plates of Mercator's atlas, and over the years improved on the Arctic map (and others) as explorers and whalers came back with ever more accurate descriptions of the coastlines, in the case of the Arctic map especially those of Spitsbergen and Nova Zembla (also, and more correctly, known as Novaya Zemlya, "New Land" in Russian).

Mercator's authoritative (but wrong) depiction of the North Pole persisted well into the seventeenth century, only to be dispelled gradually by *real* discoveries.

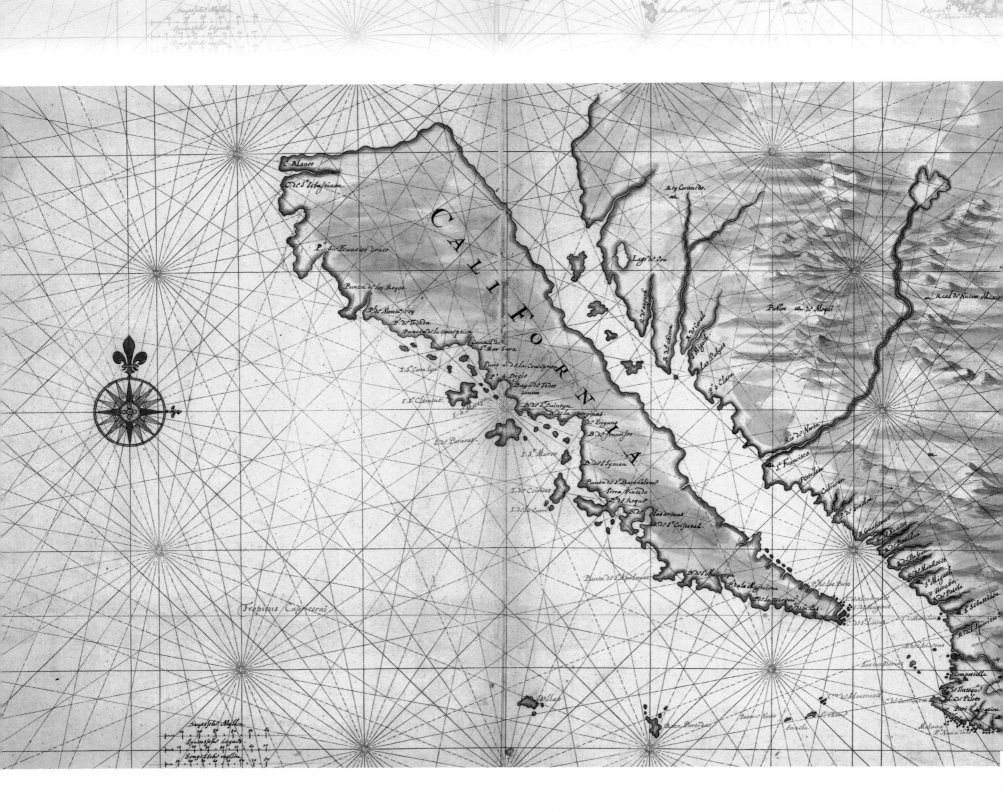

3

Black Amazon Women: The Island of California

One of the most famous misconceptions in cartographic history is of California as an island. The origin of this error is *Las sergas de Esplandían*, a romantic novel written in 1510 by Garci Rodríguez de Montalvo, stating that "on the right hand of the Indies there is an island called California very close to the side of the Terrestrial Paradise; and it is peopled by black women, without any man among them, for they live in the manner of the Amazons."

It is somehow fitting that California, now home to the entertainment industry in general and Hollywood in particular, itself should be named after a fictional place first mentioned in a novel.

Baja California, the Mexican peninsula that runs parallel to the mainland for hundreds of miles, was discovered in 1533 by a mutineer from Hernán Cortés's expedition into Mexico, followed by a trip by Cortés himself to that area (near present-day La Paz, on the southern tip of the peninsula). The lay of the land led him to believe this to be the island of "California" from Montalvo's novel.

Expeditions in 1539 and later seemed to indicate California was a peninsula, and at first it was thus shown on maps, including some by Mercator and Ortelius. Nevertheless, the idea of an insular California was revived, probably in part by the fictional accounts of Juan de Fuca. He claimed to have found a large opening in the western coast of North America, possibly the legendary Northwest Passage.

Further inspiration was the overland expedition by Juan de Oñate, who descended the Colorado River (1604–5) and believed he saw the Gulf of California continuing off to the northwest. California reappeared on the map as an island for the first time in 1622 in a map by Michiel Colijn of Amsterdam, and this image would endure far into the eighteenth century.

The idealized view of California as an insular Garden of Eden at the edge of the known world was disproved by Father Eusebio Kino's expedition from 1698 to 1701. Kino proved that Baja California is connected to it in the north. Doubts remained, however, and the issue was finally laid to rest only with the expeditions of Juan Bautista de Anza (1774–76).

Ironically, sometime in the future, California might actually become the island it was once thought to be: Tectonic activity will separate the area west of the San Andreas Fault from the mainland—although this might take thousands, if not millions, of years.

4

"We Hope the League of Nations Will Rule the Tetrahedron Well"

Notwithstanding that celestial objects of a certain mass generally are spherical in shape, an article in *My Magazine*, dated May 1918 (and titled "What the World May Come To: The School Maps as They May Be in Millions of Years to Come"), predicts that the earth is spinning itself into a tetrahedron. The explanation, as one can imagine, is very dodgy:

How would you like to live on a tetrahedron? Men say the earth will one day come to that. Many strange visions men have had of the world since it began. They used to think it a disc floating in water. There are still stupid people who believe it is flat. Every wise child knows that the truth is that the earth is practically a round ball.

It will not always be like this, however. What we really live on is an oblate spheroid; what the people of the world will live on in millions of years

to come will probably be a tetrahedron. It all looks terribly dull, but it is really tremendously interesting.

The world is now the shape of a globe, the shape which gives the biggest possible bulk for its surface, but the inside of the earth is still cooling and condensing, and the internal changes are slowly changing its shape. The surface, already condensed to its utmost, will not change with the core; it cannot reduce its area, but it adapts itself to the shrinking interior by taking a shape which occupies less bulk. So the earth is to become a tetrahedron, a sort of pyramid, the shape which gives the smallest bulk for its surface.

The article concludes:

We may be sorry for the editors and poets in those days. It is pleasant to write of sailing round the globe, or of this spinning ball, but who would

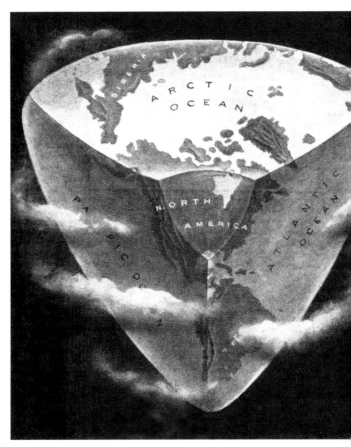

THE MAP OF THE WORLD IN THE SCHOOLROOMS OF AGES TO COME

not pity the poet who has to write and make his rhymes about some bold Sir Francis Drake's brave journey round the tetrahedron? We hope the League of Nations will rule the Tetrahedron well.

II. LITERARY CREATIONS

*The bastard children of literature and cartography, these maps
exist to illuminate the fictional worlds created by writers.*

1

More's Memento Mori: A Color Map of Utopia

A map of the world that does not include Utopia is not worth even glancing at," said Oscar Wilde. Neither can an Atlas of Strange Maps do without a map of Thomas More's fictional island, which arguably remains the most influential of all "invented countries."

Utopia was the title of a short social satire written by More in 1516, in which he contrasted the tyranny and corruption of early-sixteenth-century England with the perfect society of Utopia, a fictional island situated in the recently discovered Americas. The first of Utopia's two books is both a dialogue between the author and a traveler called Raphael Hythlodaeus. In it, More traces the ills of contemporary English society to private landownership, which he sees as the cause of mass poverty and crime. The second book portrays Utopia as a protocommunist, "utopian" society without private property or any of the concomitant so-cial calamities, and with a great diversity (and freedom) of religion.

Utopia brought More fame as instantaneous as possible in those days when books were the fastest mass medium around. This map is a color version of a woodcut made by Ambrosius Holbein for a 1518 Basel edition of More's best seller, and might have been commissioned by More's good friend the Dutch humanist Erasmus. Holbein (1494–ca. 1520), scion of the famous German painter dynasty and son and older brother to two more famous Holbeins both called Hans, modified a previous rendition of Utopia, which remained closer to the extensive topographical description of the island in the text.

The island is still clearly Utopia, with its main city Amaurotum clearly marked, as are the source (*fons*) and estuary (*ostium*) of the river Anydrus. The figure in the lower left-hand corner, pointing toward the island, is Hythlodaeus. But the overall shape of the island has changed—it now clearly represents a human skull!

It's not entirely clear why Holbein did this. Utopia certainly isn't described as skull-shaped in More's text. This is possibly Holbein's own contribution to Utopia, in the form of a memento mori. That name covers a variety of pictorial props to remind the audience of the vanity and brevity of life. The most obvious way was to portray a person holding a skull, but the Holbeins seem to have specialized in original versions. One of the most famous paintings in London's National Gallery is by Ambrosius's brother Hans Jr. On *The French Ambassadors* (1533), a seemingly incongruous diagonal streak at the foreground of the portrait turns into a skull when the painting is observed from the correct oblique angle.

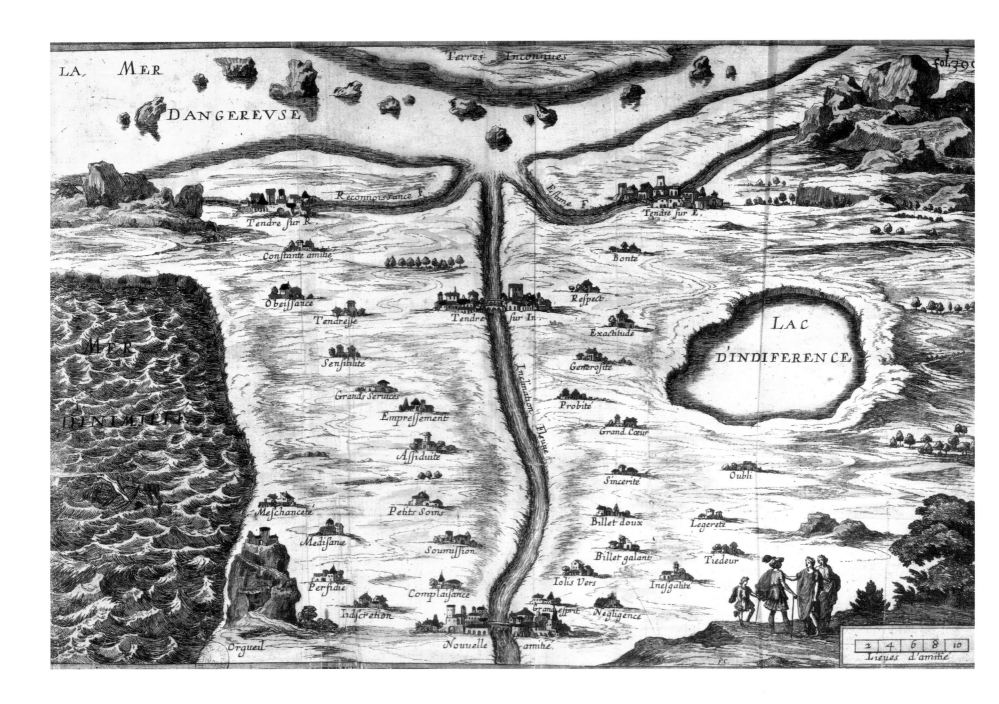

2

Love's Topography: La Carte de Tendre

One of the earliest and most influential examples of sentimental cartography is the Carte de Tendre, an example of the highly refined imagination prevalent in seventeenth-century French literary salons. The fictional country of Tendre (Tender) was inspired by *Clélie, histoire romaine*, a novel by Madeleine de Scudéry (1607–1701), whose much-frequented and tone-setting salon was one of the focal points of *préciosité*, a rarefied literary genre noted for its effusive erudition and gallantry.

The map of Tendre is a topographic allegory, representing the stations of love as if they were real paths and places. The country is bisected by the Inclination (Disposition), a river that runs south to north, joined by two smaller rivers, the Estime (Respect) and the Reconnaissance (Gratitude), before plunging into La Mer Dangereuse (the Dangerous Sea), which is separated from a reef-ridden narrows from Terres Inconnues (Unknown Lands). To the west are the decidedly choppy waters of the Mer d'Inimitié (Sea of Enmity).

The smooth flow of the rivers symbolizes the control over passions, the perils of the sea and the danger of unbridled emotions. Straddling the rivers are three eponymous capital cities: Tendre-sur-Estime, Tendre-sur-Reconnoissance and Tendre-sur-Inclination. Places along those rivers mark the waypoints of "civilized" love—and some of its pitfalls:

Marking the road from Nouvelle amitié (New Friendship) to Tendre-sur-Reconnoissance are the following towns, purportedly representing a gradual increase of affection: Complaisance (Kindness or Smugness), Soumission (Submission), Petits soins (Care of Small Things), Assiduité (Attentiveness), Empressement (Eagerness), Grands services (Great Favors), Sensibilité (Sensibility), Tendresse (Tenderness), Obéissance (Obedience) and Constante amitié (Constant Friendship).

However, close to the forbidding rock fortress of Orgueil (Pride) in the extreme southwest are places to be avoided, such as: Meschanceté (Meanness), Medisance (Disparagement), Perfidie (Betrayal) and Indiscretion (Indiscretion).

Equally avoidable are the localities leading from Nouvelle amitié toward the Lac d'Indifference (Lake Disinterest): Négligence (Negligence), Inesgalité (Inequality), Tiédeur (Lukewarmness), Légereté (Levity) and Oubli (Oblivion).

Leading toward Tendre-sur-Inclination and beyond to Tendre-sur-Estime are the towns of Grand-esprit (Great Wit), Iolis Vers (Beautiful Verse), Billet galant (Gallant Letter), Billet doux (Sentimental Letter), Sincérité (Sincerity), Grand Coeur (Magnanimity), Probité (Probity), Générosité (Generosity), Exactitude (Punctuality), Respect (Respect), and Bonté (Goodness).

Some commentators have remarked on the similarity of this imaginary landscape with that of France, in particular the area between Paris and the coast of southern England (Terres Inconnues). For others, the straight alignment of towns was reminiscent of the American Midwest. More than one noticed the similarity between the river delta and certain parts of the female anatomy.

3

Finally in Need of a Navy: New Switzerland

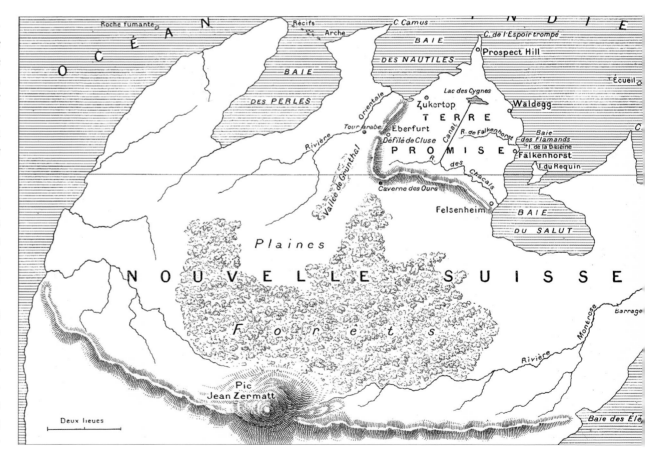

In 1900, the famous French writer of adventure stories Jules Verne published *Seconde patrie* (Second Fatherland) in two parts. Like many of the *feuilletons* at the end of his life, this adventure story was a revisiting of earlier work, though in this case not Verne's own. *Seconde patrie* is a sequel to a well-known book by Johann David Wyss: *Der schweizerische Robinson* (*The Swiss Family Robinson*).

"Of all the books I read in childhood," Verne said, "the one that I particularly adored was 'The Swiss Family Robinson.' . . . Wyss's work, abundant in events and adventures, is especially impressive to young minds. . . . I've spent years on their island! . . . How I envied their fate!"

Wyss was a Swiss pastor, who wrote this Christian morality story disguised as an adventure novel to teach his four sons about family values, self-reliance and good husbandry. It tells of a shipwrecked Swiss family's survival on an East Indian island. The family is not called Robinson, by the way; that's a reference to the earlier, equally fic-

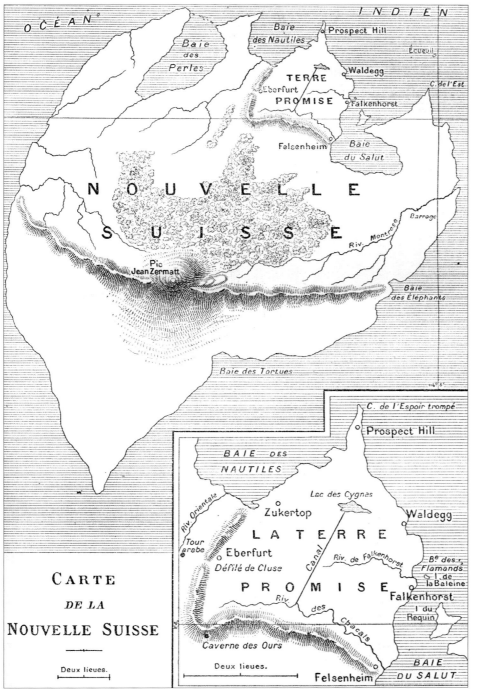

CARTE

DE LA

NOUVELLE SUISSE

Deux lieues.

tional Robinson Crusoe, probably still the most famous shipwrecked person this side of *Lost.*

Verne's book revisits the original shipwrecked family after their rescue from "New Switzerland"—but has a band of pirates seize them on their voyage to England. They are abandoned on the same island they had just escaped. Unnamed in the original, the family is called Zermatt in Verne's book. The Zermatts are joined by Jenny, a girl stranded by a different shipwreck and rescued by the Zermatt boys, and the English family Wolston. Aided by the engineering skills of Mr. Wolston, the islanders set about to further exploring and taming their new home.

Further adventures with unfriendly natives, sea travel and shipwrecks ensue. At the end, the island is annexed by Great Britain and the flourishing colony soon has over 2,000 inhabitants.

The first and second maps (in French) show Verne's take on the island: The Zermatts and their associates discover that New Switzerland is a whole lot bigger than they thought on their first stay—opening up possibilities for more adventures. . . . The third map (in German) is from a German edition of the original story by Wyss.

Flip around the German map with the flat side of the island to the south and some vague similarities with the French map will appear. The bays, islands, rivers and other named places have moved around, changed size, and so on as they only can on fictional islands. Check out these pairs:

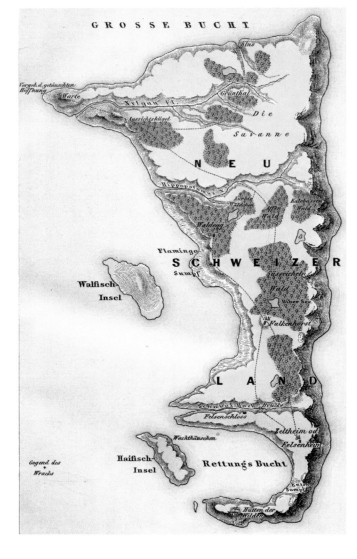

- Rettungsbucht—Baie du salut (Deliverance Bay)
- Haifisch Insel—Ile du Requin (Shark Island)
- Vorgebirge der falschen Hoffnung—Cap de l'espoir trompé (Cape of False Hope)
- Aussichtshügel—Prospect Hill
- Zeltheim oder Felsenheim—Felsenheim (German: Tent-home or Cliff-home)
- Walfisch Insel—Ile de la Baleine (Whale Island)
- Falkenhorst (Falcon's Aerie)
- Waldegg (Forest Edge)
- Schakal Bach—Riviere des Chacals (Jackal River)
- Grünthal—Vallée de Grünthal (Greendale)

Verne's map doesn't mention some other features on the original Wyss map, among others:

- Flamingo Sumpf (Flamingo Swamp), unless it transmogrified into the Lac des Cygnes (Swan Lake)
- Bibersee (Beaver Lake)
- Nylgau Fluss and Hippopotamus Fluss (Nilgai and Hippo rivers)

- Klus (a Swiss village, boasting the ruins of castle Falkenstein)
- Gegend des Wracks (area of the wreck)
- Hütten der Wilden (huts of the savages)

Verne does name other features in the "original" settlement area, cut off from the bulk of the island by an encircling mountain range:

- Terre Promise (Promised Land)
- Eberfurt (Boar Ford)
- A canal, linking Swan Lake with Jackal River
- Baie des Flamands (Flemish Bay)
- Zukertop (Sugartop)

Verne has the remarooned Zermatts discover their island is bigger than they thought, and has them discover and name, among others:

- Pic Jean Zermatt (Mount John Zermatt)
- Baie des Perles (Pearl Bay)
- Baie des Nautiles (Nautilus Bay)
- Baie des Eléphants (Elephant Bay)
- Baie des Tortues (Turtle Bay)
- Riviere Orientale (East River)
- Tour arabe (Arab Tower)
- Caverne des Ours (Bear Cavern)

4

Lord of the Flies, with a Happy Ending: A Two Years' Vacation

The mysterious island is a much-used trope in adventure stories. Long before he devised "New Switzerland," Jules Verne came up with another unknown isle for one of his stories—Chairman Island, the setting for *Deux ans de vacances* (1888).

Verne marooned a group of schoolboys sailing from New Zealand on a deserted South Pacific island and describes their struggle to stay alive. After two years on the island, they are visited by their original ship, since taken over by mutineers bent on smuggling alcohol, tobacco and firearms. The boys manage to overpower the smugglers and make their escape.

Verne's story echoes the adventures of Robinson Crusoe (Verne refers to Daniel Defoe's fictional hero in the preface), but also prefigures *Lord of the Flies,* William Golding's grimmer retelling of what happens when schoolboys are left to reinvent civilization all by themselves.

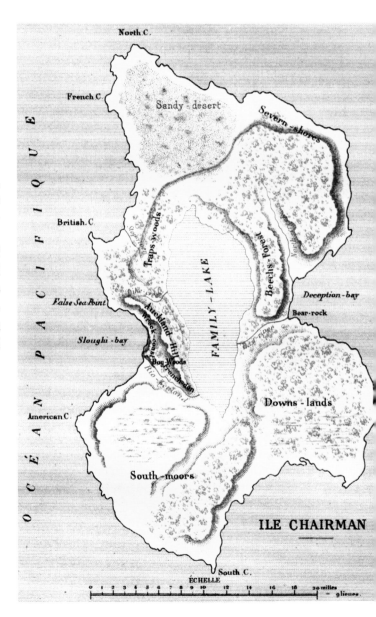

5

Captain Nemo's Deathbed: The Mysterious Island

During the siege of Richmond in the American Civil War, five prisoners of war (and the pet dog of one of them) escape via balloon and brave several days of stormy weather to crash-land on an uncharted island in the South Pacific. Thus begins Jules Verne's tale *The Mysterious Island* (1874), yet another one set on a fictional tropical island.

The five, being loyal Northerners, name the island after Lincoln. They manage to survive, thanks to the diverse talents of the group, consisting of an engineer and his man-servant, a sailor and his adoptive son and a journalist. They produce a granite house, an electric telegraph, bricks, pottery and even a seaworthy ship, getting extra help from Jupiter, a domesticated orangutan.

The mystery of Lincoln Island consists of the many unexplained rescues of the group. When they find a message in a bottle, the group set out for a nearby island, where they find Ayrton, the survivor of one of Verne's previous novels, *In Search of the Castaways*. When a group of pirates lands on the island, their ship gets blown up.

It turns out that the enigmatic force protecting the castaways is none other than Captain Nemo, who uses the mysterious island as a base for his ship. The *Nautilus* had escaped its near-certain end in the maelstrom at the close of *Twenty Thousand Leagues Under the Sea*. Nemo, the sole survivor of his crew, dies of old age and is buried at sea, in his ship.

A volcanic eruption then destroys the island; the six survivors are picked up by a ship alerted by a message sent earlier by Nemo.

Lincoln Island, vaguely octopus-shaped, bears other names referring to the Union cause in the Civil War: Baie de l'Union (Union Bay), Baie Washington (Washington Bay), Mont Franklin (Mount Franklin) and Lac Grant (Grant Lake).

Other localities are the Forêts du Far West (Far West Forests), Golfe du Requin (Shark Gulf), Cap Mandibule Nord and Sud (North and South Capes Mandible), Pointe de l'Epave (Flotsam Point), Ilôt du Salut (Deliverance Islet), Marais des Tadornes (Shelduck Swamp), Port Ballon (Balloon Harbor), Presquîle Serpentine (Serpentine Peninsula) and Promontoire du Reptile (Reptile Promontory).

All of these names add to the island's *adventurousness*, even if it's all a bit reminiscent of New Switzerland.

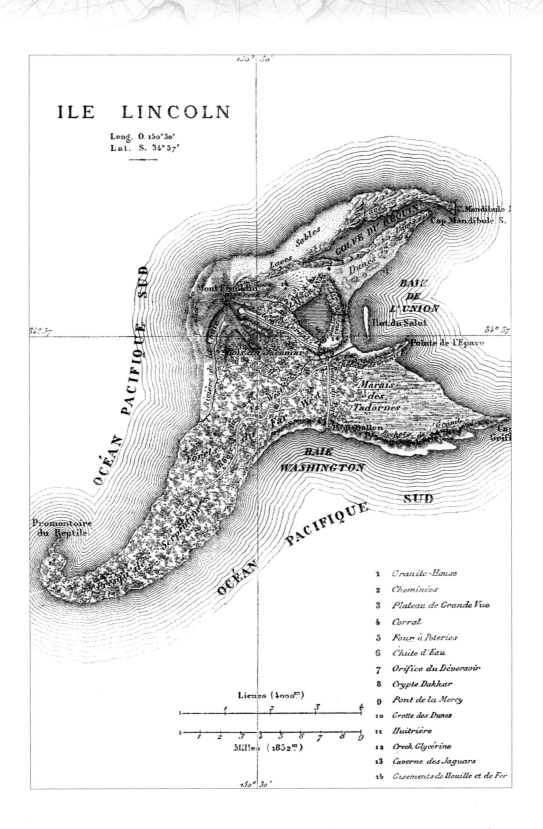

ILE LINCOLN

Long. O. 150°30'
Lat. S. 34°57'

OCÉAN PACIFIQUE SUD

Mont Franklin

GOLFE DU REQUIN

Cap Mandibule N.
Cap Mandibule S.

Laves
Sables
Dunes

BAIE
DE
L'UNION

Ilot du Salut

Pointe de l'Epave

Bois de Jacamars

Marais
des
Tadornes

Cap
Griffe

BAIE
WASHINGTON

OCÉAN PACIFIQUE SUD

Promontoire
du Reptile

Forêts

Serpentine

Presqu'île

OCÉAN PACIFIQUE SUD

Lieues (4000ᵐ)

1   2   3   4

Milles (1852ᵐ)

1 2 3 4 5 6 7 8 9

1   Granite-House
2   Cheminées
3   Plateau de Grande Vue
4   Corral
5   Four à Poteries
6   Chute d'Eau
7   Orifice du Déversoir
8   Crypte Dakkar
9   Pont de la Mercy
10   Grotte des Dunes
11   Huitrière
12   Creek Glycérine
13   Caverne des Jaguars
14   Gisements de Houille et de Fer

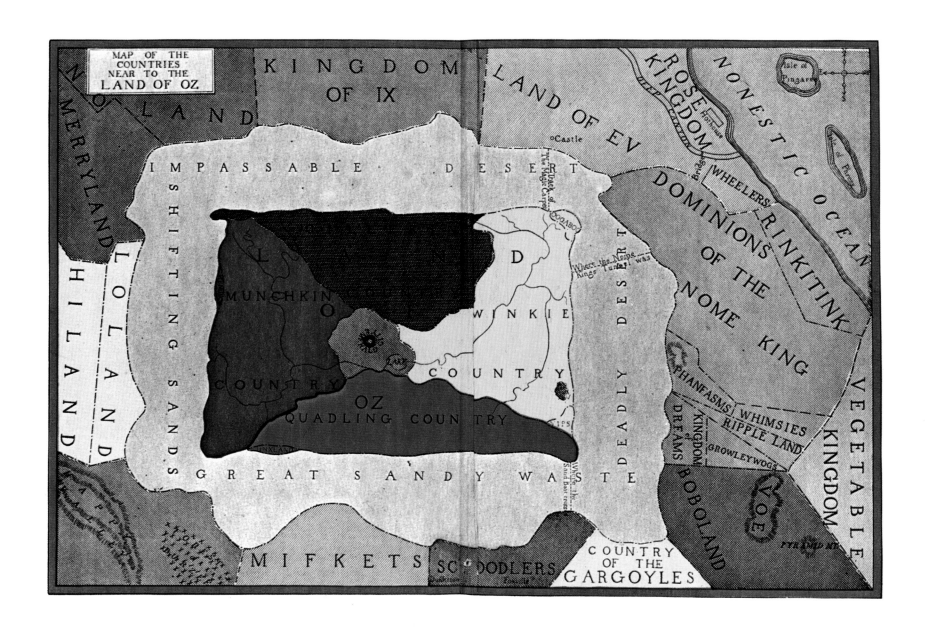

MAP OF THE
COUNTRIES
NEAR TO THE
LAND OF OZ

KINGDOM
OF IX

LAND OF EV

ROSE
KINGDOM

NONESTIC OCEAN

RINKITINK

NO LAND

MERRYLAND

IMPASSABLE DESERT

DOMINIONS
OF THE
NOME KING

WHEELERS

oCastle

HILAND LOLAND

SHIFTING SANDS

MUNCHKIN

O Z

WINKIE

COUNTRY

DEADLY DESERT

PHANFASMS WHIMSIES
RIPPLE LAND
KINGDOM OF
DREAMS
GROWLEYWOGS

VEGETABLE KINGDOM

COUNTRY

COGABO

Where the Nome
King [under] was

D

LAKE

OZ
QUADLING COUNTRY

VOE

BOBOLAND

GREAT SANDY WASTE

Where the
Sand Boat crosses

PYRAMID MT.

MIFKETS

SCOODLERS

COUNTRY
OF THE
GARGOYLES

6

Not Kansas, but Just as Square: The Land of Oz

Oz is an imaginary magical monarchy, first introduced in L. Frank Baum's book *The Wonderful Wizard of Oz* (1900). In all, Baum wrote fourteen children's books about Oz, presenting himself as the "Royal Historian" of Oz. After his death, Ruth Plumly Thompson continued the series. The books were often front- or endpapered with maps of Oz, which got more elaborate later on in the series.

The basics stayed the same, though: The Land of Oz is rectangular in shape, divided along the diagonals into four counties:

- Munchkin Country (east)
- Winkie Country (west)
- Gillikin Country (north)
- Quadling Country (south)

In the center is Emerald City, the capital and seat of Princess Ozma. Oz is completely surrounded by deserts, insulating the country from invasion and discovery. The isolation may be splendid, it is not total: Children from our world got through, as well as the Wizard of Oz and the more sinister Nome King. To prevent further incursions, the Good Witch Glinda, ruler of Quadling Country, created a barrier of invisibility around Oz.

Peculiar on some maps is that west is right and east is left (while north is still top and south bottom). Some say this is because Baum looked at the wrong side of a glass slide while copying the map. Others believe the reversed compass rose simply reflects the "confusing" nature of Oz, possibly due to Glinda's spell. The reversal of east and west makes sense in that the Wicked Witch, after enslaving the Winkies, was called the "Wicked Witch of the West"—even though Winkie County is on the right-hand side of the map. Robert A. Heinlein claims in his book *The Number of the Beast* that Oz is on a retrograde planet, spinning in the opposite direction of earth.

Oz is the largest country on the continent of Nonestica, which also includes the countries of Ev, Ix and Mo (also known as Phunniland). Nonestica lies in the Nonestic Ocean—possibly a local name for the Pacific Ocean. In fact, some hints indicate that Oz is in the South Pacific: There are palm trees and horses are non-native. In *Ozma of Oz*, Dorothy is sailing to Australia when she is washed overboard and lands on the shores of Ev. Intriguingly, Oz is commonly used to refer to Australia, which borders the South Pacific Ocean.

The origin of the word "Oz" is uncertain. One story holds that L. Frank Baum took it off a filing cabinet, which was divided into two alphabetical drawers: A–N and O–Z. Another holds that it is a corruption of Uz, the biblical homeland of Job. It could also be a reference to ounce (abbr. oz.)—with the story of Oz being an allegory for the populist struggle against the gold standard (personified by the powerless, frightened wizard in the books).

Other theories state that "Os" is an old English word for God, and in the musical *Wicked*, a clever parody on the Oz material, it is proposed that Oz derives from "oasis" or "ooze," being a reference to the creation legend of a great flood.

7

Close Your Eyes and Think of Airstrip One: The World in 1984

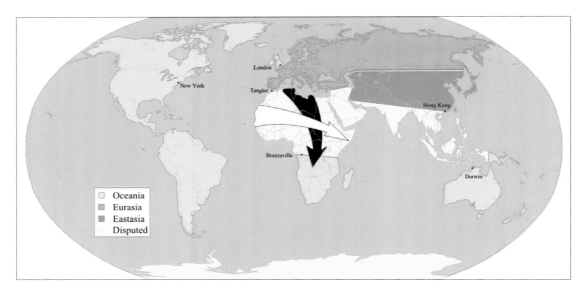

In George Orwell's dystopian novel *1984*, the world is ruled by three superstates:

- Oceania covers the entire continents of America and Oceania, the southern part of Africa and the British Isles. These are the main location for the novel, in which they are referred to as "Airstrip One."
- Eurasia covers Europe, Turkey and (more or less) the entire Soviet Union.
- Eastasia covers Japan, Korea, China, northern India and a part of Central Asia.

Unfortunately, there's not much "super" to these states except their size. All three are totalitarian dictatorships. Oceania's ideology is Ingsoc (English Socialism), Eurasia's Neo-Bolshevism and Eastasia's is the Obliteration of the Self. These ideologies are very similar, but the people are not informed of this.

The three states are in a perpetual state of warfare—sometimes two against one, sometimes all three against each other. These wars are fought in the Arctic regions and in the disputed territories, stretching from North Africa over the Middle East and southern India to Southeast Asia—in a quadrant whose corners are Tangier, Brazzaville, Darwin and Hong Kong. The black and white arrows show the Eurasian and Oceanian offensives in Africa respectively, described at the end of the book.

And yet the war might just not even be real at all. It's clear that the Oceanic media are one-sided and fabricate "facts." A dissident book central to *1984* suggests the two other powers may actually be a fabrication of the government of Oceania, which would make it the world government. Or, on the other side of the scale of thinkable alternatives: Airstrip One is not an outpost of a greater empire, but the sole territory under the command of Ingsoc, which fabricates eternal global war to keep its people permanently mobilized, scrutinized and on rations.

III. ARTOGRAPHY

The symbols and conventions of mapping can be adapted for something more artistic than cartography.

1

Back to the Drawing Board: An Inaccurate Map of Charlottesville

Back in the days when cartography was based on hearsay instead of satellites, mapmakers stressed that their work was the best available by calling them "accurate" maps (which often still left a lot to be desired).

Nowadays, we can rely on those satellites and other high-tech measuring equipment to deliver us maps without phantom islands, mysterious monsters lurking in the seas, or missing continents. Which means it's again okay to go completely nuts in the overly neglected field of subjective mapping.

Russell Richards is an artist living and working in Charlottesville, Virginia. Much of his work is cartography-inspired. Very little of it would be handy if you got lost. This "inaccurate map" of Charlottesville is a deliberate reversal of the "accurate" maps of yore: the fabulation of a real landscape through the distorting lens of memory.

Richards painted a series of Charlottesville maps upon returning to his childhood home after a twelve-year absence. "It's about the places that have been important to me," he says, "some of which are no longer there, and the memories they evoke."

Charlottesville is a small (pop. 40,000) city in central Virginia, and pleasant enough to be elected "best place to live" in 2004. It was the home of Presidents Jefferson, Monroe and Madison. Jefferson's Monticello mansion is a major tourist destination. Jefferson also founded the University of Virginia, still residing in Charlottesville.

2

Drawn from Memory: United Shapes of America

This canvas doesn't depict a herd of abstract cows, but a collection of maps. Artist Kim Dingle asked teenaged schoolkids in Las Vegas to draw the shape of their country from memory. She collated their sketches into these "United Shapes of America"—one of the curious maps in *You Are Here*, the excellent collection of unconventional maps by Katharine Harmon.

Are there any general conclusions to be drawn from these drawings? Well, yes:

- Alaska and Hawaii are missing in all the maps (as they often are in depictions of the United States, which must be quite frustrating for Hawaiians and Alaskans), but maybe the kids were told to disregard these outlying states.

- Almost every map captures Florida's peninsularity—although in some maps, the Sunshine State is better endowed than in others.

- The second most prominent protrusion is New England's, jutting north into Canada by way of Maine; also represented quite frequently is Texas's southern extremity, sometimes rendered as one of America's two south-oriented jaws, the other being Florida. That these two features are less "popular" than Florida might be down to the fact that they're land borders and don't show up prominently on maps that include the landmasses of Canada and Mexico.

- Some students add another extremity in the west, probably the Baja California peninsula—non-U.S. but unmissable on any map including that part of Mexico.

- The Great Lakes' indentation in the U.S.'s northern border is quite difficult to render from memory (give it a try!) and therefore generally ignored or minimized.

- A few manage to get the bulging of the U.S. West Coast, but many draw a straight line there; in fact, a lot of these drawings resemble not much more than box shapes.

- Some shapes deviate wildly from the box-shape standard: One map in the center of the canvas seems to depict the state of Nevada (in which the students live). Another, near the bottom, has notched edges, giving it the appearance of a postage stamp more than anything else.

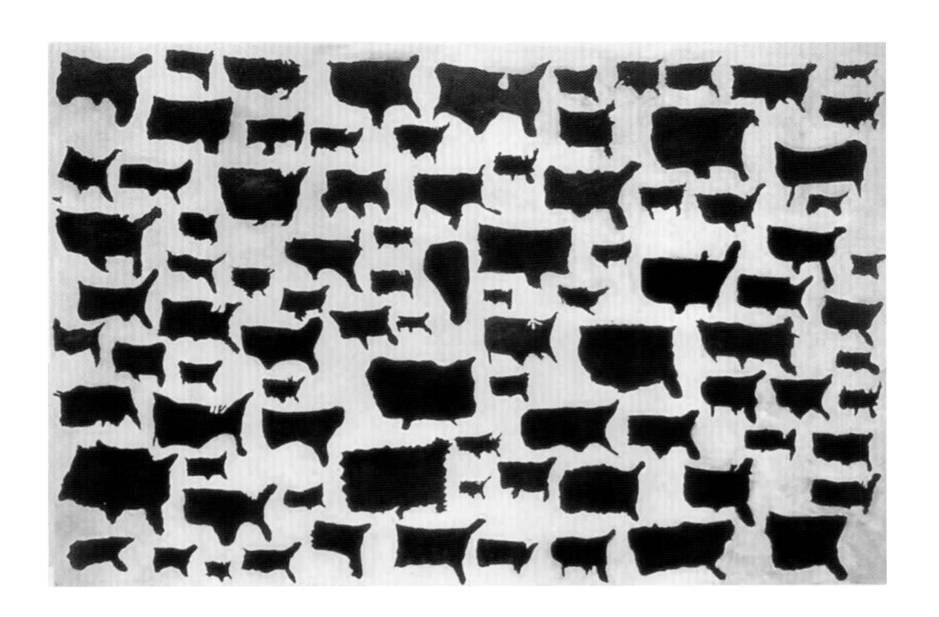

3

Gothic Barcelona: Horror and Humor in El Born

Medieval beginnings and a bustling nightlife combine to make El Born (The Source) one of the best-known "gothic" haunts of Barcelona. In the daytime, go to the Picasso Museum in the Calle de Montcada, visit the imposing cathedral of Santa Maria, or go shopping in the trendiest boutiques of Barcelona. At night, step out to sample the local cava or the many cocktails on offer in the numerous bars and clubs of the neighborhood, centered along the Passeig de Born.

Born is about a ten-minute walk away from the beach, and another ten-minute walk from the Ramblas and Barcelona's city center. The nearest metro stop is Jaume 1. This map of El Born is taken from *A Weird and Wonderful Guide to Barcelona* from a series of city guides by Le Cool, a publishing house that takes pride in producing weird and wonderful maps for their guidebooks.

Each map reflects the character of the area portrayed, including this one, says Andrew Losowsky, editorial director of Le Cool: "The Born is a Gothic part of town, with trendy bars and shops mixed in with dark corners and overgrown windowboxes. Old meets new in a trendy, grungey kind of way."

An ominous, gothic spookiness is summoned on this map by the somber choice of colors—black, gray, blue—the dark silhouettes "standing" and "sitting" on the street grid, and the way the streets branch out into eerie roots, leaves and vines. The streetlamp and the candlestick only emphasize the dusk settling in around them. A butterfly is poised to flutter into a spider's web. The grotesque sometimes turns into the comical: a giant faucet filling up a glass of wine, a street turning into a guitar, tentacles gripping designer shoes . . .

4

Now This Is World Music: Harmonious World Beat

Now this is world music: a map of the planet's continents translated into the notes, ties, bars and staves of traditional sheet music. According to some musically more talented commentators on the blog, this song opens on a strong C major chord, with an F major 7 appearing in central Russia and China, which is then repeated around the Great Lakes. If played at the indicated tempo, it would last about forty seconds.

And it doesn't sound as unharmonious as you would expect. We are familiar with the shape of the earth's continents, and they look just about "right" on this stunning map. But that's a visual impression; those shapes are essentially jagged, capricious and random. It doesn't necessarily follow that their outline would make good music. And yet somebody ran these notes through a musical program, and the result doesn't sound half bad.

"Cheers to [James] Plakovic for the extra effort that makes this something quite clever rather than a mere novelty," wrote one commentator. "To me, the notes south of the Cape of Good Hope sound like winds, blowing strongly where oceans meet," offered another. "If arranged correctly, it could become a very good piano piece," concludes someone else. "At least I think so."

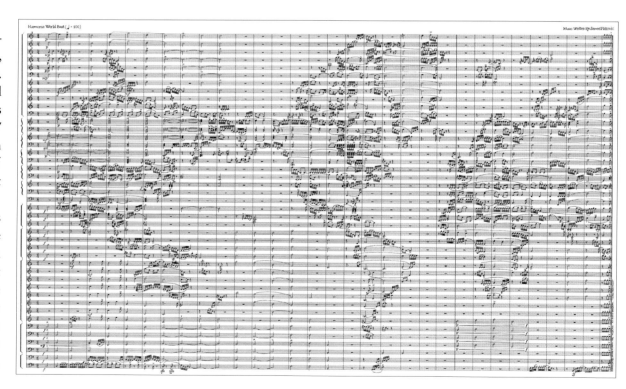

5

Chimero's Chimeras: Transformed States of America

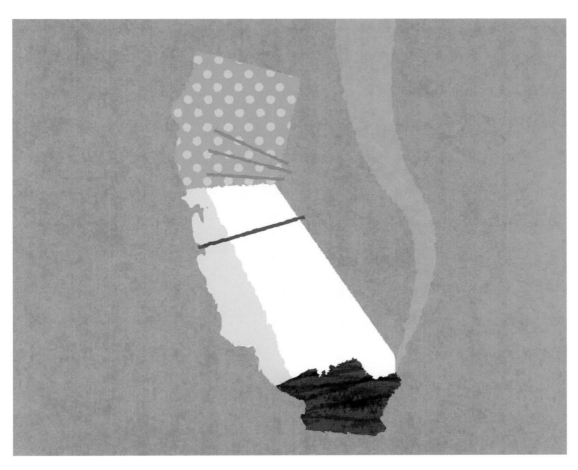

Frank Chimero always thought that California had the shape of a stubbed-out cigarette. Which would have remained a non sequitur, had he not been a graphic designer. "After some discussion and a quick sketch, I decided that I could sort of make a game for myself out of doing the same exercise with the other 49 states," Chimero says on the Web page where he exhibits an ongoing project simply called The States, and which is, he promises "no content, just form."

The way Chimero fills out each state's borders is not necessarily related to any aspect of the state's history, culture or whatever. But sometimes they do, and other times . . . it's tempting to read a connection into Chimero's peculiar state maps.

Take for instance the California Stub, its bent shape a familiar sight in many an ashtray—except that you'd be harder pressed to find an ashtray in health-conscious California than in most other states. Since 1994, a statewide smoking ban has effectively excluded all smoking from workplaces, restaurants and

bars and outside areas less than 20 feet (6 m) from any window, door or air intake of any government building in the state. Since 2008, it's a misdemeanor to smoke in a car in California in the presence of a minor. Some cities have imposed even stricter limits on outdoor smoking, for example forbidding it in public parks (as is the case in Los Angeles). If California weren't already so smoke-free, Chimero's cigarette might have been the best logo for a campaign at making it so.

An owl might be a bit more difficult to link to Illinois, apart from the fact that the state's shape lends itself particularly well to strigiform impersonation. The state is home to eight different species, including the short-eared, the long-eared and the eastern screech owls. The Illinois Owl, Chimero reveals, also refers to an owl family of the endangered, long-eared variety that in the beginning of 2008 chose to build their winter nest on Chicago's South Side, in the South Loop.

6

Pretentious, Moo? The Cow That Conquered the World

Cows are as quintessentially Dutch as clogs, tulips and windmills. The most typical breed of Dutch dairy cow is the black-and-white variety, which has been perfected in the northern part of the Netherlands for the better part of two millennia.

These cows have been exported the world over, notably to North America, where they are known as Holsteins. Their slightly divergent relatives in Holland are known as Friesians. Holsteins were reimported in Europe to be crossbred with Friesians.

This cow reflects the internationalist bent of the Holstein/Friesian: Its black spots are in the shape of countries, no doubt each with significant numbers of originally Dutch livestock roaming its pastures—Ireland and the boot of Italy at its hindquarters, the Netherlands in the middle and Great Britain and France to the right. Other spots might be parts of other countries. Or, to paraphrase Freud, sometimes maybe spots are just . . . spots.

7

Don't Do It, Baby: Giant Infant Threatens Downtown Vancouver

Despite having given the world Boris Karloff and Bryan Adams, Vancouver is consistently ranked as one of the world's best cities to live in. Canada's West Coast metropolis (2.25 million inhabitants) owes its bustle to the arrival of the Transcontinental Railway (1887) and the opening of the Panama Canal (1914), combining to make it Canada's busiest seaport. Vancouver gets its sparkle from the subsequent diversification of the economy into high tech, now also including a sizable movie industry,

giving it the nickname "Hollywood North."

"I love Vancouver," says New York illustrator Aaron Meshon, who produced this disconcerting painting for a *Vancouver Magazine* article on the city's baby fair. "I really wanted to include the whole city in a Godzilla-inspired scene." Meshon avoided the clichés of helicopters and tanks attacking the baby, lending the scene an eerie calm. The Vancouver Baby, in a little bear hat, stands in the waters of Coal Harbour, stretching out its hands

toward downtown Vancouver, connected to the south of the city via (west to east) Burrard, Granville and Cambie bridges. Next to Cambie Bridge is BC Place Stadium, the world's largest air-supported domed stadium.

To the baby's back, on the tip of the peninsula it shares with downtown, lies Stanley Park, one of the largest city parks in North America. On the far shore lie West and North Vancouver, and farther north the North Shore Mountains.

8

"Not Atoll": Real Maps Reassembled into Imaginary Places

Maps are man-made interpretations of the physical world around us, in order for us better to grasp and navigate it. Turn that definition on its head, and you may arrive at something like this—a map that has the look and feel of "real" topography, but is nothing more (or less) than an artist's vision that has no pendant in the real world.

"I have become increasingly fascinated by the intersection of man-made and natural forms made visible in maps and atlases," says Francesca Berrini, the Portland, Oregon, artist who produced this work and other map-based art, which explores "the colorful geometry of political divisions laid over the organic forms of the continents. . . . While human boundaries and routes of travel shift and vary, features of the landscape seem to remain solid underneath the flow of humanity."

Well, landscape does move because continents do shift, but the pace is so glacial that we hardly notice it. That just won't do for Berrini, who applies scissors and glue to vintage maps to speed up and manipulate the tectonic process. The resulting collages of islands, archipelagoes and continents are new lands, formed where her imagination collides with those map fragments.

This work is entitled *Not Atoll*, a name that might be construed as a reference to other artfully imagined, nonexistent places such as Utopia ("No Place") and Erewhon ("Nowhere" spelled backward—oh well, almost). The collage of square upon vintage map-square emphasizes the work's constructed, Frankenstein-like nature, even in the two dimensions represented here. At the same time, Not Atoll's islands reassuringly have the random shapes we have come to expect of real ones. "Real" islands, mind you. Even if the maps that represent them are equally man-made . . .

9

Maps as Art: Painted Peru

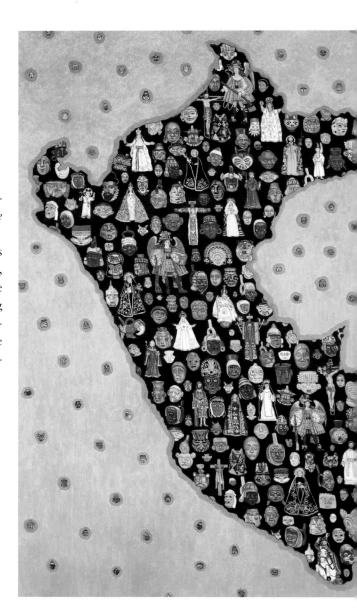

Christa Dichgans (born 1940 in Berlin) is a German painter who has shown a proclivity toward cartography in her work. Generally, her map paintings consist of a monochrome background surmounted with the contours of a country or continent, filled up with stuff of all sorts.

Amerika represents a gazillion dollar bills forming the lower forty-eight states, contrasting with a yellow background. *Europa* shows a continent consisting of an uncountable number of human faces against a blue background. Other examples of this technique are *Deutschland*, *Indien*, *Der schwarze Kontinent* and this oil painting, *Peru*.

This canvas shows the contours of this South American country. Inside its borders, small religious and folkloric figurines are pasted over the black background, creating the impression of something like a cigar label collection. Outside Peru's borders, the yellow background is embroidered with buttonlike embellishments.

IV. ZOOMORPHIC MAPS

*For reasons varying from humorous to mnemonic, mapmakers
often transform their maps into living creatures.*

SCOTT'S GREAT SNAKE.

Entered according to act of Congress in the year 1861 by J. B. Elliott of Cincinnati in the Clerks Office of the District Court of the Southern District of Ohio.

1

Scott's Great Snake; or, The Anaconda Plan

We propose a powerful movement down the Mississippi to the Ocean, with a cordon of posts at proper points . . . the object being to clear out and keep open this great line of communication in connection with the strict blockade of the seaboard, so as to envelop the insurgent States and bring them to terms with less bloodshed than by any other plan.

—General in Chief Winfield Scott, in a letter to Major General George B. McClellan, dated May 3, 1861

The American Civil War lasted from 1861 to 1865 and cost over 600,000 lives—about ten times the American death toll in Vietnam, and still the highest number of U.S. casualties in any conflict. If U.S. Army (i.e., Northern) general in chief Winfield Scott (1786–1866) had had his way, the number of casualties might have been a lot lower. At the beginning of the war, he devised a plan that would have ended the secession of the Southern states with minimal loss of life.

This plan involved strangling the Southern economy by a twofold blockade. On one hand, the north would impose an economic blockade of Southern seaports, preventing the export of cash crops such as tobacco and cotton and the import of arms. On the other hand, the Union would take control of the Mississippi River, thus dividing the main part of the Confederate States of America from its westernmost parts on the right bank of the river.

The popular newspaper cartoon pictured here gave Scott's scheme its name: the "Anaconda Plan," after the giant snake that throttles its victims. The name anaconda is borne by four types of South American snake, which makes the etymology even more paradoxical. For the consensus is that the name originates in faraway Sri Lanka, but it's doubted whether it is Sinhalese ("thunder snake") or Tamil ("elephant killer") in origin.

Scott's plan was not well received; the public mood called for a large-scale invasion. President Lincoln didn't choose: He implemented the blockade as proposed by Scott, *and* the large-scale invasion. A total of two million Union soldiers repeatedly tried to capture Richmond, the CSA capital in Virginia, contributing to the eventual heavy toll in lives.

This map, dating from the beginning of the Civil War, is illuminated with several symbols and slogans, starting from the snake's tail:

- In the state of New York is placed a winged helmet, the attribute of Hermes (Mercury), the god of trade. The helmet is labeled "Free Trade"; a figure seemingly armed with a bayonet is rushing south.
- Pennsylvania is the location of what appear to be ordinary farmhouses.
- In New Jersey, the capital, Trenton, is marked out—in the wrong place—with dates referring to its important role in the events of (17)76, and (18)61.

- Maryland is labeled "We give in," while in Virginia, the western, pro-Union part of the state is about to secede (West Virginia was recognized as a state in 1863). The snake's tail is coiled around a flagpole planted at Washington, D.C., crowned with a phrygian cap, symbol of freedom since the French Revolution.
- In Virginia, a beehive crowned with a Confederate flag is seen discharging its inhabitants all over the South.
- In North Carolina, a person is scooping up rosin (a form of resin obtained from pines), clarified by the slogan "Poor eating." A string of small stick figures is seen escaping from South Carolina; a larger runaway slave, headed north with his knapsack, is labeled "Contraband."
- Strangely, the figures run toward the "Knocksville Whig," the name of a paper, published in Knoxville, Tennessee, that was pro-slavery despite also being pro-Union.
- In Georgia, a disused cotton factory is falling into ruin, due to the economic blockade. In northern Florida, a black stick figure in a patch of green might signify a runaway slave hiding in the woods and/or swamps of the region.
- A disgruntled Alabaman complains, "Dam old Virginia, took our capitol." Montgomery, Alabama, functioned as the Confederate capital for a few months in 1861, before it was replaced by Richmond, Virginia. "Burning Massa out," in neighboring Mississippi,

probably refers to local slave risings prompted by the war and its prospect of freedom.
- "A Union Man" with a noose round his neck shows how dangerous it must have been to show your sympathy for the North's cause in Louisiana. "Can't ship now," says a figure to his companions, idly lying on cargo, untransportable due to the blockade.
- A Texan firing on runaway slaves is reminded, "Costly shooting $1000.00 a head."
- Indian Territory (later to become Oklahoma) is taken up by an Indian smoking a pipe, wigwams in the background and a baby hanging from a tree branch by its wrapping.
- In the state of Kanzas (sic), "Union Music" can be heard.
- The head of the snake is chasing after "Jackson & Co." on the Missouri-Arkansas border, observed by what appears to be an Arkansas militiaman (with two daggers drawn, sporting the slogan "Hold Me") and a Tennessean with a telescope. From southern Illinois, a Union cannon is trained on the Rebels.
- Kentucky is quite literally sitting on the fence, labeled "Armed Nutrality" (sic); Illinois is illustrated with a plant, named "U.S. sucker"—the state's nickname used to be "the Sucker State," possible after the tobacco plant ubiquitous in its southern half.
- Indiana contains an illustration of a

pork barrel, and a train headed east to Ohio, also covered by sheaves of grain, illustrating the respective agricultural riches of both states.
- Iowa is endowed with marksmen bearing the "Hawk-Eye" epithet still current for the state (but dating from an earlier reference to an eponymous scout in *The Last of the Mohicans*, published in 1826).

Winfield Scott was also known as "Old Fuss and Feathers" and the "Grand Old Man of the Army." Here are ten things you—probably—didn't know about this interesting warhorse:

1. He was an active-duty general for over forty-seven years, longer than any other person in American history, serving under fourteen presidents from Jefferson to Lincoln and commanding soldiers in five wars: the War of 1812, the Mexican-American War, the Black Hawk War, the Second Seminole War and the American Civil War.

2. During the Mexican-American War, Scott commanded the southern army, landing at Veracruz and (on purpose) following the same route to Mexico City as Hernán Cortés in 1519.

3. Fat and vain, Scott was haunted by a quote from a letter from Mexico to the secretary of war that was published to sabotage his reputation. "At about 6 PM as I sat down to take a hasty plate

of soup" became a catchphrase that appeared in cartoons and folk songs for the rest of his life.

4. After the Mexican War, he served as military governor of Mexico City. He was nominated for U.S. president by the Whig Party in 1852, but lost to Franklin Pierce. He was promoted to lieutenant general in 1856, the first American to hold that post since George Washington.

5. During the War of 1812, he urged that British POWs be executed as retaliation for the Brits' executing thirteen Irish-American POWs whom they considered their own subjects, and therefore traitors. President James Madison refused.

6. He earned his nickname "Old Fuss and Feathers" for his insistence on discipline and decorum in the U.S. Army, at that time mostly a volunteer force.

7. In 1839, he helped defuse the territorial dispute between Maine (United States) and New Brunswick (Britain), which caused the so-called Bloodless Aroostook War. In 1859, he traveled to the Northwest to settle another faux conflict with the British over San Juan Island, the so-called Pig War.

8. Scott translated several Napoleonic manuals into English, including *Infantry Tactics*, which was the standard drill manual for the U.S. Army from 1840 to 1855.

9. The phrase "Great Scott!"—an interjection akin to present-day favorite "Oh my God!"—may refer to him, as in his later years he weighed three hundred pounds.

10. Winfield Scott is not to be confused with Winfield Scott Hancock (1824–1886), who also served with distinction in the Mexican-American War, also was a Union general during the Civil War, and also ran unsuccessfully for president afterward (defeated by Republican James Garfield in 1880). "Hancock the Superb" was in fact named after the other Winfield Scott, by then already famous as a hero of the War of 1812. And the latter was the commander of the former during the Mexican War. Another Winfield Scott is the songwriter who wrote the song "Return to Sender" for the eponymous Elvis Presley movie.

2

Itching to Spread Its Wings West: The American Dove, 1833

When this map was published in 1833, the United States didn't have an East Coast yet, for lack of a West Coast. The gigantic Louisiana Territory, acquired some thirty years earlier from the French, gave America dominion over the Mississippi basin, but Mexican land and the Oregon Territory, claimed by Great Britain, still stood between the United States and its "Manifest Destiny"—to stretch "from sea to shining sea."

That's a line from Katharine Lee Bates's song "America the Beautiful," composed in 1893 when the West was won, mainly by the territorial gains made following the Mexican-American War of 1846–48. It would be many decades before all the lands between Mississippi and Pacific would enter the Union as full-fledged states, but by mid-nineteenth century the iconic, instantly recognizable shape of America's lower forty-eight states was born.

Back in 1833, this mapmaker tried to attach America's iconic bird to the shape of the nation at that time. The result was this Eagle Map of the United States, Engraved for Rudiments of National Knowledge.

The map represents America as an eagle (although it looks more like a dove), its head coinciding with New England (except Maine), its eye punching a hole in Vermont, its neckline following Lakes Ontario and Erie, the wing outlining Lakes Huron and Superior (and farther west the eventual Canadian-American border at the 49th parallel). The eagle's breast follows the Atlantic seaboard, its talons form Florida—even though the claws protrude far from the coastline, and somewhat ominously, toward Cuba.

The real reason why this particular iconic representation of America's national bird never caught on is in the tailfeathers—shaped to follow a border no longer in existence after 1848. Only the western borders of the subsequent independent and later U.S. state of Texas are still instantly recognizable, in 1833 still the dividing line between the United States and Mexico. The feathers follow the U.S. inland border as it moves north, and disappears out of sight at the area disputed with Great Britain. Hardly any part of this watershed border has been preserved in the straight lines making up most of the boundaries of the western states.

Meanwhile the great inland empire of Louisiana was already being divided up; the territory was renamed Missouri Territory to avoid confusion when Louisiana achieved statehood in 1812. Strangely, no other name was chosen for the remaining territory after Missouri itself became a state in 1821 (although areas east of the Missouri River became part of Michigan Territory in 1833, and part of Wisconsin Territory in 1836, when Michigan became a state).

This map was published in Philadelphia in 1833 by Carey & Hart, in a now extremely rare atlas, the *Rudiments of National Knowledge, Presented to the Youth of the United States, and to Enquiring Foreigners, by a Citizen of Pennsylvania.*

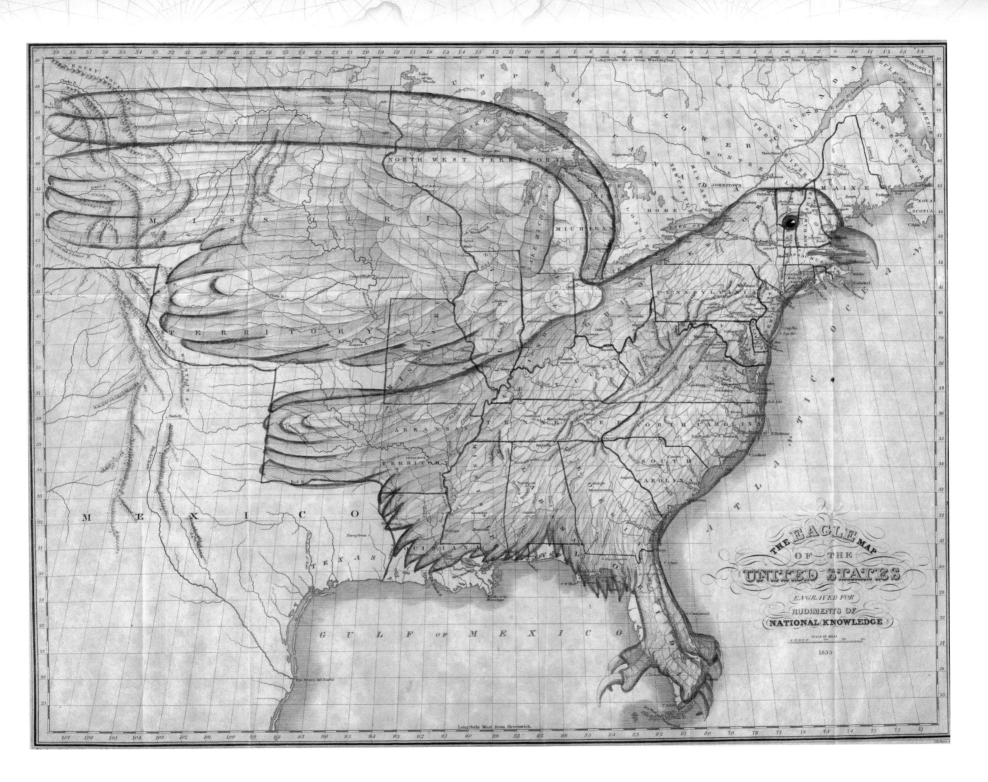

THE EAGLE MAP
OF THE
UNITED STATES
ENGRAVED FOR
RUDIMENTS OF
NATIONAL KNOWLEDGE
1833

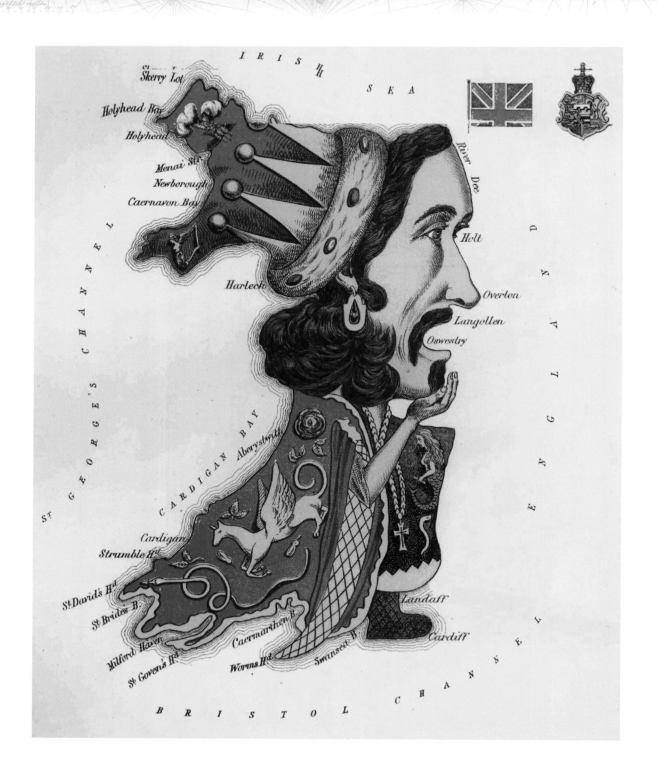

IRISH SEA

Skerry Lot

Holyhead Bay

Holyhead

Menai Str

Newborough

Caernarvon Bay

ST GEORGE'S CHANNEL

CARDIGAN BAY

Aberystwith

Cardigan

Strumble Hd

St David's Hd

St Brides B.

Milford Haven

St Goven's Hd

Caermarthen B.

Worms Hd

Swansea B.

BRISTOL CHANNEL

Harlech

River Dee

Holt

Overton

Langollen

Oswestry

ENGLAND

Landaff

Cardiff

3

Geographical Fun: The Aleph Maps

The year 1869 saw the publication in London of a very peculiar sort of atlas, *Geographical Fun: Being Humourous Outlines of Various Countries*. The book showed twelve anthropomorphous depictions of equally many European nations, each figure dressed in appropriately typical garb, all stretching and crouching to twist their bodies into a shape resembling the outer borders of their respective countries. Stereotypes were not shunned: Denmark was presented as a female ice skater, Russia as a bear, Scotland as a "gallant piper," Ireland as a peasant woman with child, and so on.

"It is believed that illustrations of Geography may be rendered educational," stated the writer, "Aleph" (pseudonym of Dr. William Harvey), in the introduction, "and prove of service to young Scholars who commonly think Globes and Maps but wearisome aids to Knowledge. . . . If these geographical puzzlers excite the mirth of children, the amusement of the moment may lead to the profitable curiosity of youthful students and imbue the mind with a healthful taste for foreign lands."

While the rhymes accompanying each map were Harvey's, the maps themselves were the product of Lillian Lancaster, née Eliza Jane Lancaster and better known perhaps as a pantomime artist both in Britain and the United States, best remembered for performing the song "Lardy Dah" onstage in New York (the origin of the still current expression "La-di-dah"). In the preface, Aleph tells of how she drew the humorous maps at age fifteen to entertain her bedridden brother. Back in Britain, Ms. Lancaster married and ended her stage career, after her husband's death moving to Brighton, where she produced many more satirical maps.

WALES

Geography bewitch'd—Owen Glendowr,
In Bardic grandeur, looks from shore to
   shore,
And sings King Arthur's long, long pedigree,
And cheese and leeks, and knights of high
   degree.

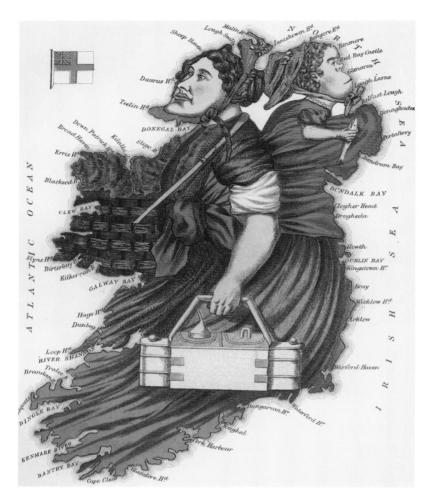

GERMANY

Lo! Studious Germany, in her delight,
At coming glories, shewn by second sight,
And on her visioned future proudly glancing,
Her joy expresses by a lady dancing.

DENMARK

For Shakespeare's Prince, and the Princess
of Wales,
To England dear. Her royal spirit quails;
From skating faint, she rests upon the snow;
Shrinking from unclean beasts that grin
below.

IRELAND

What shall typify the Emerald Isle?
A Peasant, happy in her baby's smile?
No fortune her's,—though rich in native
grace,—
Herrings, potatoes, and a joyous face.

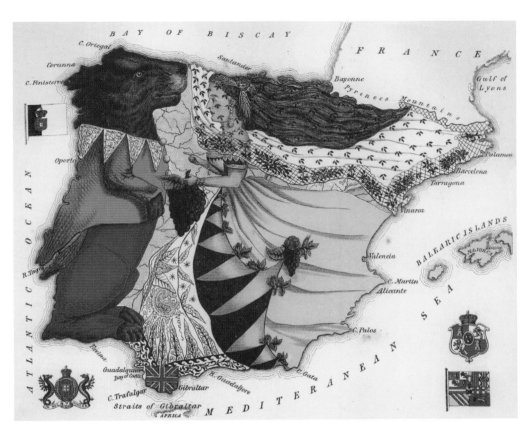

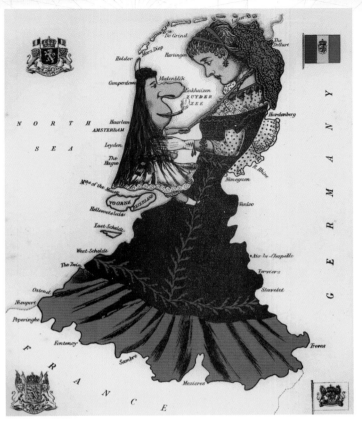

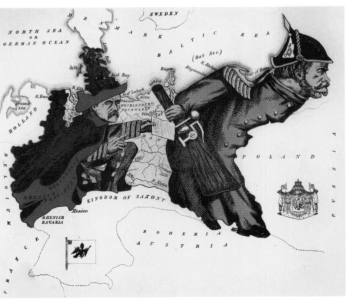

HOLLAND AND BELGIUM

Dame Holland, trick'd out in her gala
    clothes,
And Master Belgium, with a punchy
    nose;
Seem on the map to represent a land,
By patriot worth, and perfect art made
    grand.

PRUSSIA

His Majesty of Prussia—grim and old—
Sadowa's King—by needle guns made
    bold;

With Bismark of the royal conscience,
    keeper,
In dreams political none wiser—deeper.

SPAIN AND PORTUGAL

These long divided nations soon may be,
By Prim's grace, joined in lasting amity.
And ladies fair—if King Fernando rules,
Grow grapes in peace, and fatten their
    pet mules.

V. (POLITICAL) PARODY

*Here are some samples of parodic cartography—an undervalued subsection
of the ancient and distinguished art of political cartooning.*

1

A "New Species of Monster": The Gerry-mander

That will do for a salamander," proposed the painter Gilbert Stuart to the editor of the *Boston Gazette* one day in early 1812. Stuart had just added wings, claws and a monstrous head to a map showing the contorted shape of a newly formed electoral district. The editor had a better name for the monster snaking its way through Essex County in eastern Massachusetts: "*Gerry*-mander!"

For when the *Boston Gazette* published the map on March 26, mock-scientifically explaining the etymology of this "new species of monster, which appeared in Essex South District in Jan. 1812," it was clear to everyone that it was an attack on Elbridge Thomas Gerry (1744–1814), the Democratic-Republican governor of Massachusetts, who had approved the improbably shaped district to spite his Federalist opponents in upcoming elections.

Gerry lost the subsequent gubernatorial elections—in part because the sobriquet "Gerry-mander" swiftly became part of the parlance of the times, shackling his name to a practice seen as an odious example of dis-

honest party politics. If Gerry's name is remembered today, it is because redrawing the boundaries of electoral constituencies for political gain—or gerrymandering—is still current to this day.

All of which is a bit unfair to poor old Elbridge Gerry, and not just because posterity chose to mispronounce his surname (which rhymes with *Gary*, not *Jerry*). The Massachusetts politician has a few other claims to fame: He was a signatory of the Declaration of Independence (1776) and the Articles of Confederation (1778), but was one of three delegates who refused to sign the Constitution (1787), as it did not include a Bill of Rights.

Last but not least: Although he approved it as governor, the redistricting idea was not Gerry's, and in private he even opposed it. After losing the governorship, Gerry became the fifth U.S. vice president (in 1813, under President James Madison) and gained a final, sad distinction by becoming the second VP to die in office. The curse of the Gerry-mander?

History gave the artist who created the Gerry-mander a better deal. Gilbert Stuart (1755–1828) was one of the foremost portraitists of the young Republic, painting around a thousand of the most prominent Americans of the time, including the first six U.S. presidents. Most of us even own one of his works: Stuart's unfinished portrait of George Washington (1796) graces the obverse of the U.S. one-dollar bill.

2

The Jesusland Map

I only thought it was worth a few laughs," says Gavin Webb of the Jesusland map he created on November 3, 2004, the day after George W. Bush's reelection. He was wrong. Jesusland traveled cyberspace like wildfire, popped up on thousands of Web sites and spawned numerous spin-off maps. Webb—*nomen est omen*—had created an Internet meme.

The satirical map was a big hit in the offline world as well: Jesusland was discussed on TV programs across America and the world, got printed on T-shirts and coffee mugs and even inspired a pop song by the same name. Why did this map, created on a whim, get such long legs?

Despite its simplicity, Jesusland succinctly and succesfully transmits several messages—objective electoral results, the depth of political division in the United States, an emotional reaction to "four more years" of Bush and a blend of satire, despair and resignation. For this reason it is, in my humble opinion, one of the greatest cartographic cartoons ever.

The map distinguishes the states that voted for Republican incumbent George W. Bush from the ones that went for John F. Kerry, his unsuccesful Democratic challenger. Kerry won a handful of populous states in three disparate places: on the West Coast, in the Midwest and the Northeast. Bush was victorious in a contiguous area covering the rest of the lower forty-eight, plus Alaska.

This geopolitical dichotomy isn't new—conservatives disparage the West Coast as the "Left Coast," liberals snobbily refer to the middle of the country as "flyover country." Bush's 2004 victory seemed to have widened the rift. It left many Democrats in despair, even ashamed. They had been able to dismiss Bush's contested victory in 2000 as a fluke, but this time he had won the popular vote, albeit with a flimsy 50.7 percent to Kerry's 48.3 percent. To many, the United States now felt like two countries.

The fact that all the areas where Kerry won are adjacent to Canada (conveniently forgetting Hawaii) must have given Mr. Webb his idea. Canada is generally perceived as more liberal than the United States in general, or in this case, more in tune with "Kerryland." It could be a haven for all those despondent liberal states, all of them merging into a "United States of Canada." Bush's election was due in large part to the massive block of religious-conservative voters, prompting Mr. Webb to baptize their country "Jesusland."

This name would not be self-applied by its inhabitants—in fact, it's a putdown toward them from the liberal camp, equivalent to the "flyover" sneer. In other versions of the map, "Jesusland" is sometimes labeled "Redneckistan." That it became so popular after the election may be because it explains the Democratic defeat by pointing the finger at the religious right and its unwavering, unquestioning support for Bush.

One variant of the map expands Jesusland into Canada by annexing the province of Alberta, noted for its more conservative voting patterns.

3

Flanders Sleeps with the Fishes: Wallonie-sur-Mer

The little kingdom of Belgium has always been a difficult proposition. It is one country shared, with increasing unease, between two peoples: the Dutch-speaking Flemings (in the north, about 60 percent of the population) and the French-speaking Walloons (in the south, almost 40 percent).

As with fault lines in geologically active zones, this divide is an irritant in the best of times, and in moments of crisis threatens to tear the country apart. It took Belgian politicians on both sides of the divide almost a year to agree on a new government following the elections of June 10, 2007. The prospect of a Belgian divorce seems to be off the table for now, but the gap between north and south shows no sign of closing.

In fact, both communities have drifted ever further apart since the start of a gradual process of federalization in the early 1970s. To offset the oppositional nature of this essentially binary federation, the Belgian federal system is extremely intricate, balancing territorial and personal definitions of citizenship. The result is no fewer than six governments:

- one for Wallonia (dealing with territorial aspects such as land management);
- one for the French-speaking community (competent in personal matters such as education and health care);
- one for Flanders (which merged the personal and territorial competences);
- one for Brussels (territorial; as Brussels is officially bilingual, the Flemish and Francophone governments both exert their personal competences there);
- one for the tiny German-speaking minority in the east of Wallonia (personal, since the Walloon government is territorial);
- and the federal government, in principle to be composed of equally many Francophone as Flemish ministers. The prime minister, although necessarily either Dutch- or French-speaking, is considered as a linguistic eunuch.

Flemish politicians insist on delegating more power to the regional level, allowing Flanders many of the institutional tools that "real" states have—and in the process disentangling them from the "solidarity" with the economically deprived south of the country. Francophone politicians see this as creeping separatism and vehemently oppose any reform that could be seen as damaging the Belgian state—or Francophone interests.

The resulting gridlock for some observers indicates that Belgium has reached the end of its tether, which is not an improbable proposition—were it not for Brussels. The capital of Belgium isn't just also the capital of Europe: It's also the capital of Flanders, which maintains its parliament there. But 85 percent of the *bruxellois* are Francophone, and thus not inclined to think kindly of incorporation into Flanders. Annexation by Wallonia is rather impractical, as Brussels is completely surrounded by Flemish territory.

Maybe an international task force of chess grand masters and Nobel Prize winners un-

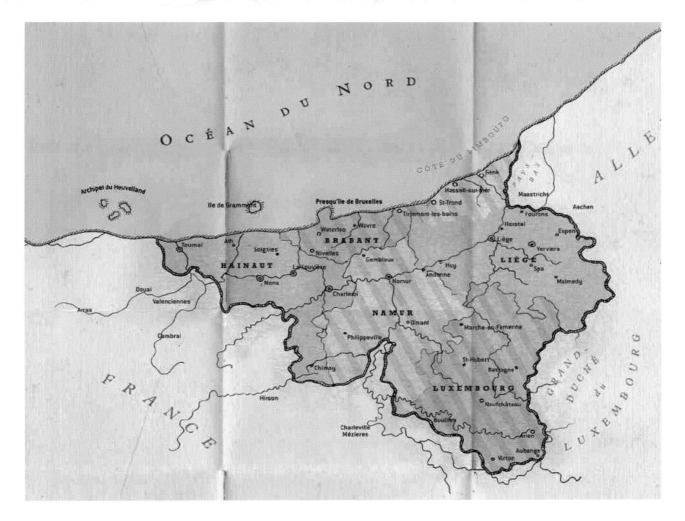

der the auspices of the Dalai Lama could find a solution everybody could live with. Barring that, another possibility would be to let nature take its course.

This map outlines the latter solution to the Belgian conundrum: Just wait until almost all of low-lying Flanders is submerged, leaving only some of its higher parts above water, i.e., the Heuvelland (Hill Country), the Ile de Grammont (Grammont being the French name for the Flemish town of Geraardsbergen)

and bits of Limburg, the Flemish province farthest from the coast (with "Hasselt-sur-Mer") and Flemish Brabant ("Tirlemont-les-Bains"; Tirlemont is the French name for Tienen). Brussels is not submerged, but connected to Wallonia via a land corridor—demanded by some Francophone politicians, as it would end the territorial isolation of the Francophone *bruxellois*, but obviously vehemently opposed by Flemish politicians.

This map, giving landlocked Wallonia ac-

cess to the North Sea, was published during the aforementioned institutional crisis in 2007 by the Francophone Belgian newspaper *Le Soir* under the title "Bientôt, on ne devra plus parler de séparatisme" (Soon, we won't need to discuss [Flemish] separatism anymore). It apparently followed comments by a Francophone politician that solidarity within Belgium works both ways, and that Wallonia would accommodate Flemings if and when their part of the country was flooded by rising sea levels.

4

A New Simplified Map of London

"Why, Sir, you find no man, at all intellectual, who is willing to leave London. No, Sir, when a man is tired of London, he is tired of life; for there is in London all that life can afford." Thus spoke Dr. Johnson in 1777.

Nowadays, it's more a question of, who can afford life in London? True, the British capital offers just about every experience known to man or woman (except having a drink in a bar after eleven at night—legal, but still a rarity), but the costs of housing and living are excruciating.

This creates two Londons: One where the very rich can still live as if they were middle class, and another where everyone is a loser, even those with an income high enough to qualify as middle class or over anywhere else.

Which seems to be the sardonic and much more succinctly put message of this new, simplified and hilarious map of London—drawn by a connoisseur of the city, by the way. The rich area on this map corresponds quite well with some of London's better areas: Kensington, Chelsea, Notting Hill.

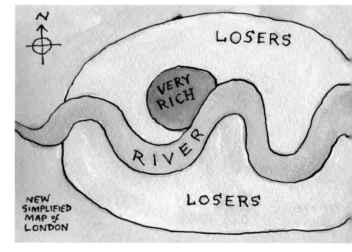

VI. MAPS AS PROPAGANDA

These cartographic projections and exaggerations are meant to scare, cajole and energize—many of them involving, in some way, Germany.

1

Overlapping Claims : Early-Twentieth-Century Balkan Aspirations

"There was hardly any part of the territory of Turkey in Europe which was not claimed by at least two competitors," said a report by the International Commission to Inquire into the Causes and Conduct of the Balkan Wars in 1914, accompanying this map.

At the beginning of the twentieth century, the Ottoman Empire was fast losing its grip on its Balkan possessions. However, as the emerging Balkan states had overlapping claims to the newly liberated territories, tensions ensued, causing conflicts and wars that continue to cast dark shadows over the region.

In 1912–13, the first of two Balkan Wars saw a Balkan League, composed of Bulgaria, Montenegro, Serbia and Greece, conquer Macedonia, Albania and most of Thrace; the second Balkan War was a fight between the victors over those spoils. Such conflicts might have been difficult to avoid, due to the fragmented distribution (i.e., "balkanization") of the different ethnicities, and the absence of well-established borders after centuries of Ottoman rule.

- Serbia (on the map still called "Servia") was a small, recently independent statelet smaller even than present-day Serbia, yet it championed the rights of all southern Slavic peoples, striving to extend its reach from all the way up to Trieste—deep into Austro-Hungarian territory— down to Montenegro, then already an independent state.
- Bulgaria, also newly independent, similarly aspired to create a "Greater Bulgaria" by annexing all of ancient Macedonia, including the (now Greek) port city of Salonika.
- Romania (then Rumania) wanted to extend its reach to include areas with large numbers of ethnic Romanians: Bessarabia (now the former Soviet republic of Moldavia) and Transylvania (then part of Austria-Hungary).
- Greece controlled only the Peloponnese, some islands, mainly in the Aegean Sea, and a small part of the mainland to the north (including Athens). It sought to

extend its domain to Crete, Cyprus, most of the Aegean islands, the southern half of Albania and Macedonia, and Thrace, up to and including Constantinople, now the Turkish city of Istanbul.

None of these four countries managed to hold on to the maximum extent of their territorial aspirations as shown on this map. Even after the wars that followed the breakup of Yugoslavia in the 1990s and Kosovo's unilateral declaration of independence in 2008, there's still enough irredentist frustration in the Balkans to at least ensure that local history will never be a merely academic subject.

2

Germany Wins World War I: French Worst-Case Scenario

Is this what Europe would have looked like had Germany won World War I? It is, according to this map, published in the French newsmagazine *Sur le vif* right after the end of the conflict in 1918. The caption beneath the map reads, "The dream of German hegemony—What defeat would have cost us."

"We publish here, after the *Daily Mail*, one of the maps summarizing the German plans to rearrange Europe after the war, which shows what would have become of us if the Germans would have won.

"Belgium, Luxembourg, Serbia and Montenegro would have disappeared completely. Russia would have lost all its Baltic provinces and Petrograd [St. Petersburg]. Great Britain would have become an Austro-German colony. As for France, it would have been reduced to the *département des Basses-Pyrénées*, with Bayonne as its capital."

The map itself paints an even starker picture than the indignant caption. Germany and Austria-Hungary have divided most of Europe among themselves, leaving just a handful of nations on the European mainland independent: the Netherlands, Denmark, Switzerland and Spain (neutral throughout the war); that Italy and Romania are left untouched on this map may indicate that this plan was drawn up in the first year of the war, for both countries, initially neutral, entered the war on the side of the Allies (Italy in 1915, Romania in 1916). Paradoxically, Bulgaria, an ally of Germany and Austria-Hungary, is nevertheless absorbed by the Austro-Hungarians.

Grande Allemagne (Greater Germany) encompasses all of France (except the aforementioned "reservation" around Bayonne), Belgium, Luxembourg, the Baltic states and a large part of the Russian Empire. All of Britain is labeled *colonie allemande* (German colony). Austria-Hungary has acquired all of the Balkans (except Romania) and, apparently, Russia up to and including Moscow—plus Ireland.

Whatever the true origin of this map—French, British or German—the reason for publishing it after the war seems to be to convey the message: This is what we fought against. It is very doubtful whether this plan would have been executable after a victory of the Central Powers. Austria-Hungary was already teetering on the brink before the war, being the explosive mix of nationalities that it was. To think that it would have been able to extend its control over such a large swath of the Balkans and eastern Europe—not to mention Ireland—seems altogether fanciful. German plans did call for annexation of coastal and industrially important French regions, but never as far south as the Pyrenees.

It would be useful to reverse Cicero's famous adage (*cui bono*, i.e., "to whose benefit?") and ask, in looking for the culprit: *cui malo?* France and Great Britain are both obliterated in this plan, so it might very well be this map is propaganda of either French of British fabrication.

3

Another Plan to Kill France: Italy on the Atlantic

Europe without France—not even a tiny French rump state at the foot of the Pyrenees: that was the dream of Major Adolf Sommerfeld, who developed his plans in a work bluntly entitled *Frankreichs Ende* (The End of France).

Sommerfeld's thesis circulated during the First World War, which pitted Germany and France against each other, to an extremely lethal effect. Both countries had been continental archenemies for some time, and one way to stop the constant threat of war would be to obliterate the enemy completely.

Interestingly, this Pan-German Scheme for the Extinction of France shows no German expansion toward the east, even if this *Drang nach Osten* (Eastward Urge) had been a staple of German geopolitics since the nineteenth century, and would inform the Nazis' war with the Soviet Union a generation later.

Germany's westward expansion includes the northern two-thirds of France, from the Atlantic coast just north of the Garonne estuary to the Swiss border at Geneva. Bel-gium, whose neutrality had been violated by the German invasion of August 1914, would remain independent and, intriguingly, the British would get a bridgehead on the Continent: a bit of northern France between the Somme and the Belgian border.

Major Sommerfeld's solution for France's southern part is puzzling—he hands it to a country that fought the First World War on the Allies' side: Italy gets Corsica and the French mainland, from Nice to Biarritz, all the way north to the newly Italian city of Bordeaux—gondolas soon bobbing on the Garonne River . . .

4

A Patchwork Germany to Keep France Safe

Whereas German plans to rearrange Europe after a World War I victory wiped France off the map, French post-victory plans were more generous: There would be not just one Germany, but half a dozen.

This map represents A Popular French Conception of Reconstructed Europe after defeating Germany, and was first published as "L'Europe future" in 1916. It shows a longing for the time before German unification, when the German states were too small and divided to pose a threat to any of their neighbors.

A victorious France would only annex the Palatinate, a small part of Germany north of Alsace-Lorraine. Belgium would acquire German territory bounded by the rivers Moselle and Rhine, with Bonn straddling the German-Belgian border. An elongated neutral zone running from the Dutch to the Swiss borders would shield these territorial gains from the rest of Germany.

What remained of Germany would be divided into several small independent states: Westphalia, up to the river Weser; Hanover, covering much of northwestern Germany; Saxony in central Germany; Wurtemburg and Bavaria in Germany's south; Prussia, also covering Silesia, from the Baltic coast to independent Bohemia; and an Austria extended to incorporate German-speaking populations in Moravia (and retaining an exit to the Adriatic via parts of Slovenia and Croatia).

The rest of Austria-Hungary would be divided among Hungary, Rumania and a greater Serbia, which would also include Albania. Montenegro would be enlarged also. Bulgaria would retain its access to the Aegean via eastern Thrace. The Balkan borders seem a lot more sketchy than those in Germany, indicating which area was more crucial to French thinking about a postwar "reconstruction" of Europe.

5

Germany's Future: From the English Channel to the Caspian Sea

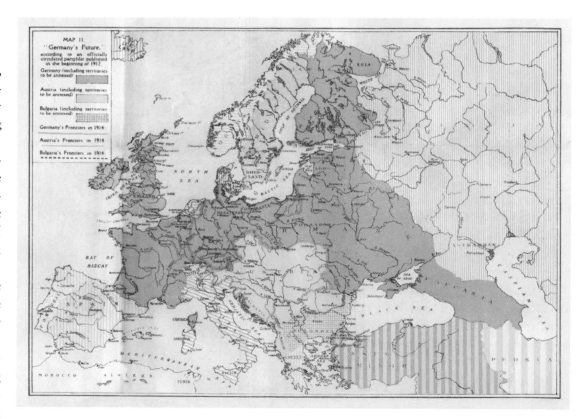

ontrary to the previous map on page 60, this one shows a radically different Germany, victorious after World War I—only minimally expanding westward, directing most of its energy toward the east.

Germany does annex Belgium and northern France, but these acquisitions are dwarfed by the enormous land grab in central and eastern Europe: Poland, the Baltic states, Belarus, Ukraine and a large part of western Russia, all the way up to the Kola peninsula, plus Finland.

Eerily, this roughly corresponds to the farthest extent of the Nazi inroads into the Soviet Union, two and a half decades after the end of hostilities in World War I.

Germany's allies also expand: Austria-Hungary grabs land to the south, annexing Albania, Serbia and a large part of Romania. Bulgaria annexes Dobrudzha (from Romania), Slavic Macedonia and southern Serbia.

All this leaves just a handful of states on the European mainland, apart from Germany, Austria-Hungary and Bulgaria: Turkey, Greece, Italy, Switzerland, France, Spain, Portugal, Norway, Sweden and the Netherlands, its southern panhandle awkwardly sticking into the newly greater Germany.

6

What a German Africa Might Have Looked Like

Germany achieved unification only in 1871, urgently desiring all the trappings of a modern European state, which in the nineteenth century included a colonial empire. Frustratingly, most of the "uncivilized" world had already been carved up by others, notably Great Britain and France.

Imperial Germany nonetheless sought to find its own "place in the sun" in the few places left on earth that were still open for colonization—which in the late nineteenth century mainly meant Africa. From 1883 onward, the German Empire managed to rapidly acquire a string of colonial possessions in Africa and Oceania. Germany's far-flung territories were all handed over to other powers following the Treaty of Versailles (1919).

In the Pacific, Germany managed to take control of the northern half of present-day Papua New Guinea and a string of island groups, most administered as part of Deutsch-Neuguinea (German New Guinea). Germany's main African colonies were:

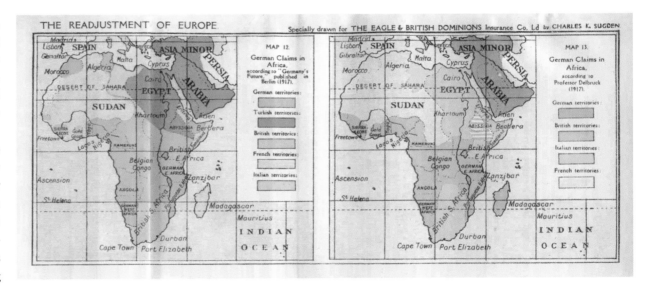

- Deutsch-Südwestafrika (German South West Africa) was mandated by the League of Nations to South Africa, which administered it as South West Africa until it became independent in 1990 as Namibia—the only former German colony to still house a significant German minority.

- Deutsch-Ostafrika (German East Africa) consisted of Tanganyika and Ruanda-Urundi. The former became a British mandate and united in 1964 with Zanzibar to form Tanzania, the latter became a Belgian mandate and in the early 1960s two independent nations—Rwanda and Burundi.

- Deutsch-Westafrika (German West Africa) made up of Kamerun and Togoland; both colonies were divided between France and Great Britain. British Togoland later joined Ghana, the French part became independent Togo. Part of British Cameroon later joined Nigeria, the other part joined French Cameroon in independence.

Obviously, had Germany won World War I, it would have retained its African colonies—and annexed the colonies of the defeated powers. Two plans painted a slightly different picture of what this Deutschafrika (German Africa) would have looked like.

- A 1917 plan called "Deutschlands Zukunft" (Germany's Future) would have seen the emergence of a contiguous German colonial empire covering Central Africa by annexing colonies from Portugal (Angola and Mozambique), France (Congo-Brazzaville), Belgium (Congo-Léopoldville) and Britain (Nigeria, Rhodesia, British East Africa). Separate German colonies would be established in Morocco, Algeria and Somalia ("Berbera"). The Ottoman Empire—a German ally—would retain control over Egypt and large parts of Sudan and Abyssinia (Ethiopia). The Italians would get Libya, Eritrea and Somaliland. The French would be allowed to retain "their" Sudan (i.e., the northwestern part of Africa), while the British would keep South Africa, Sierra Leone and the Gold Coast.

- A second, less ambitious plan by a Professor Delbruck (1917) would eliminate German "Berbera" and Germany's North African holdings, allow the French to keep more of their Saharan and Central African empire and the British to keep their East African possessions, as well as Egypt and Sudan, and more of southern Africa. Germany's colonies in West Africa would cover more ground, though, up to and including the Gold Coast, with a separate territory from Sierra Leone to Senegal. In this plan, Madagascar would also have become German.

7

Reverse-Engineered Nazi Propaganda: "Germany Must Perish"

German people! Now you know what your eternal enemies and opponents have thought out for you—against their plans for your extermination, there is only one possible course of action: fight, work, win!" Those lines are a sample of the Nazi propaganda surrounding this map, showing, as its legend reads, "possible dissection of Germany and apportionment of its territory." The map is real enough, but it does not show, as the German text below states, "the carve-up of Germany and its apportionment to other states, *as planned by world plutocracy* [emphasis added]."

Adolf Hitler once said that a lie had to be big and bold in order to be more easily believed by the masses. This big, bold lie has some truth mixed in, to make it even more believable. This map was indeed designed by a Theodore N. Kaufman (as indicated by the copyright notice in the lower right-hand corner).

Kaufman, a theater ticket agent in Newark, New Jersey, self-published *Germany Must Perish* in March 1941, when Germany

had already overrun much of Europe (but the United States still hadn't entered the war). In the hundred-page pamphlet, Kaufman advocated the sterilization of all Germans and the total annexation of Germany by its neighbors.

The incandescent rage in Kaufman's book might be co-explained by the fact that Kaufman was Jewish, and that even at that point, before the *Endlösung* was set in motion, the Jews had sufffered a great deal under the Nazis, first only in Germany, now also in the rest of occupied Europe. But Kaufman's rage backfired spectacularly. The Nazis reverse-engineered his booklet for their own propaganda needs and milked it for maximum effect.

For starters, they presented Mr. Kaufman as a "typical" Jew, referring to him as *Nathan*

Kaufman (the N. in Theodore N. Kaufman actually stands for Newman). *Germany Must Perish* was a worst seller, and had no social or political impact whatsoever in the United States. But the Nazis presented it as essential to "plutocratic" (i.e., American and British) political thinking, suggesting that Kaufman was a close adviser to president

Franklin Delano Roosevelt. FDR probably had never even heard of Kaufman before the German propaganda kicked in. The Nazis presented Kaufman as the president of a so-called American Federation of Peace, while in fact he is not known to have been a member of any Jewish or anti-German organization. He was, to use an appropriate German term, your typical *Einzelgänger*.

American journalist Howard K. Smith was in Germany when the Nazis got their hands on Kaufman's book, which "provided the Nazis with one of the best light artillery pieces they have, for, used as the Nazis used it, it served to bolster up that terror which forces Germans who dislike the Nazis to support, fight and die to keep Nazism alive."

Germany Must Perish was used by the Nazis as an excuse for further discriminations against the Jews. After the war, revisionist historians argued that the Holocaust was a response to Jewish plans for the extermination of Germany. Ironic how evildoers always end up accusing their opponents of planning the very crimes they perpetrate themselves.

Kaufman's map shows prewar borders (in thin lines) and his proposal for postwar borders (in bold lines):

- Poland did in fact move westward after the war, but not as far as this map shows.
- The Czechs contented themselves with regaining the Sudetenland after the war, and didn't march into Austria, Saxony and Silesia, as shown in this map.
- Even the neutral Swiss seem to be grabbing their little piece of Germany and prewar Austria.
- And the French gobble up most of southern Germany, up to and including the entire state of Bavaria, making France a neighbor of "Czechia."
- Belgium seems to use the postwar confusion to annex Luxembourg and southern parts of the Netherlands, as well as a band of German territory along its border.
- The greatest winner would appear to be Holland (i.e., the Netherlands), which would gain the most territory relative to its prewar size by annexing most of northern Germany, giving it the seaports of Bremen and Hamburg, and access to the Baltic.
- Denmark would move south a bit, annexing most of Schleswig-Holstein.

VII. OBSCURE PROPOSALS

Being a collection of plans for states that never left the drawing board, or at least haven't done so yet.

1

Dampieria to Guelphia: A Ten-State Australia

This map, dated 1838, shows an early proposal for the subdivision of Australia into ten states. It was published by the *Journal of the Royal Geographical Society* in London, and accompanied an article entitled "Considerations on the Political Geography and Geographical Nomenclature of Australia." In it, the following divisions were proposed, but never enacted:

- Dampieria: northwestern Australia (named after the seventeenth-century captain William Dampier, the first Englishman to explore and map parts of Australia).
- Victoria: southwestern Australia—far from the present-day state of Victoria in the southeast (after Queen Victoria).
- Tasmania: part of present-day Western Australia and Northern Territory—not the present-day island state (named after the Dutch explorer Abel Tasman, the first European known to have reached Tasmania—in this context, Van Diemen's Land—and New Zealand).

- Nuytsland: near the Nullarbor Plain (after Pieter Nuyts, Dutch governor of Formosa and explorer; Nuytsland is still the name of a nature reserve in Western Australia).
- Carpentaria: south of the Gulf of Carpentaria (named after Pieter de Carpentier, a governor-general of the Dutch East Indies).
- Torresia: northern Queensland (named after Luis Vaez de Torres, the Spanish explorer who first reported that Australia and New Guinea were separated, by what came to be known as the Torres Strait).
- Cooksland: near Brisbane, in New South Wales and Queensland (named after the British captain James Cook, first recorded European to reach Australia's east coast).
- Guelphia: present-day Victoria, most of New South Wales, part of South Australia (named after the House of Guelph, an originally German dynasty that was the origin of many British monarchs).

- Van Diemen's Land: what is now Tasmania (named after Antonie van Diemen, governor-general of the Dutch East Indies, by Abel Tasman).
- Flindersland: after captain Matthew Flinders, early circumnavigator and explorer of Australia (a name he helped popularize); his name is attached to more than one hundred places and geographical features in Australia, though not to one of its states.

That strange blob of land south of the Victoria/Nuytsland coast? That's Portugal and Spain, lifted from their spot between the Atlantic and the Mediterranean, twisted around and dropped here, for size comparison purposes. The Iberian Peninsula is located here as far south of the equator as its actual position is north of the equator, thus demonstrating that Australia is as close to the equator as is North Africa.

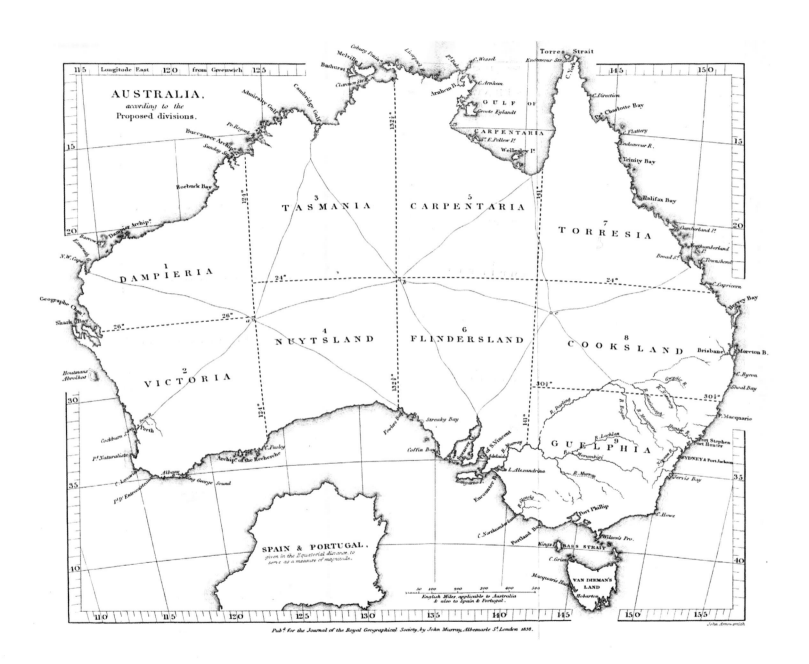

AUSTRALIA,
according to the
Proposed divisions.

SPAIN & PORTUGAL.
given in the Equatorial distance, to
serve as a measure of magnitude.

English Miles applicable to Australia
& also to Spain & Portugal.

Pub.d for the Journal of the Royal Geographical Society, by John Murray, Albemarle St. London 1838.

2

"That Absurd Element in Jefferson's Mind": Ten States That Never Were

Thomas Jefferson's plan for the division of the Northwest Territory into ten new states never left the drawing board, and no one seems to regret it. The proposed names were just too silly, writes nineteenth-century Jefferson biographer John T. Morse Jr.:

The names suggested for these ten States are a peculiar mixture of Latin and Indian, and while a semblance of some of the names still remains in two cases [i.e., Michigania and Illinoia], in all others it is so absolutely forgotten that the very fact has ceased to be known by many close students of American history. Yet, besides this humane and noble piece of statesmanship [the proposed prohibition of slavery in the territory] we have a glimpse of that absurd element in Jefferson's mind which his admirers sought to excuse by calling him a "philosopher." The matter is small, to be sure, but suggestive. He proposed as names for the several

subdivisions of this territory: Sylvania, Michigania, Cheronesus, Assenisippis, Metropotamia, Illinoia, Saratoga, Washington, Polypotamia, and Pelipsia.

In early 1784, a committee chaired by Thomas Jefferson presented a draft of what was to become the Northwest Ordinance, whereby the fledgling United States for the first time expanded beyond the borders of the original thirteen states to incorporate the area between the Great Lakes and the Mississippi and Ohio rivers.

Apart from the provision that "after the year 1800 of the christian aera (*sic*), there shall be neither slavery nor involuntary servitude in any of the said states, otherwise than in punishment of crimes" (deleted from the draft at the instigation of southern delegates, but reintroduced in the final ordinance), Jefferson's proposal is notable in that it stipulated for the first time that new states would be created out of the new territories, thus resolving the long-simmering rivalries between

THOMAS JEFFERSON'S *Conception* FOR THE SUBDIVISION OF THE NEW WEST. THIS PROPOSAL WAS CONTEMPORARY WITH HIS ORDINANCE OF 1784 FOR THE GOVERNMENT OF NORTHWEST TERRITORY, BUT WHICH NEVER BECAME EFFECTIVE

Except for the classical names for states, note how prophetic the plan Jefferson had never seen the West.

the original states, many of which had overlapping claims out west.

Virginia, Connecticut, Massachusetts, North and South Carolina and Georgia all successively relinquished their claims to territory between the Appalachians and the Mississippi to the new federal government.

Jefferson's proposed states were to be two degrees latitude in width, where possible, and arranged in three tiers, one between the Mississippi and a meridian through the falls of the Ohio River (the only navigational obstacle in the river, close to Louisville, now in Kentucky), a second between this meridian and one through the confluence of the Great Kanawha and Ohio rivers (at Point Pleasant in West Virginia, of *Mothman Prophecies* fame) and a third, smallest tier between the second meridian and the Pennsylvania West Line.

The faux-antique state names, derived from Latin, Greek and Native American languages, were to be:

In the western tier:

- Sylvania (from the 45th parallel to the Lake of the Woods) would have covered much of present-day Minnesota, Michigan's Upper Peninsula and some of northern Wisconsin. Its name refers to the dense hickory, pine and oak forests of the area.
- Michigania (from the 43rd to the 45th parallel) would have stretched from Lake Michigan to the Mississippi River, incorporating most of Wisconsin and a slice of western Michigan.

- Assenisipia (from the 41st to the 43rd parallel), named after the Assenisipi River, now better known as the Rock River, would have mainly covered the northern part of modern-day Illinois. Its name was sometimes rendered as Assenisippis or Assenisippia.
- Illinoia (from the 39th to the 41st parallel), covering a large southern part of the present-day state of Illinois, named after the Illinois River.
- Polypotamia (from the 37th to the 39th parallel), or "Land of Many Rivers," so named because the waters of the Wabash, Ohio, Missouri, Illinois, Mississippi and other rivers intermingle there.

In the middle tier:

- Cherronesus (north of the 43rd parallel), sometimes also spelled Chersonesus (Greek for "peninsula"), would have covered most of Michigan's Lower Peninsula and part of its Upper Peninsula.
- Metropotamia (from the 41st to the 43rd parallel), or "Mother of Rivers," from the belief that this area was the source of many rivers.
- Saratoga (from the 39th to the 41st parallel), whose name celebrated the Revolutionary War victory of the same name, incorporated a large part of Indiana and some of Ohio.
- Pelisipia (from the 37th to the 39th parallel), named after Pelisipi, a native name for the Ohio River.

In the eastern tier:

- Washington, hemmed in by the second-tier states, the Ohio River, Lake Erie and Pennsylvania, corresponding to the eastern part of Ohio.

Whereas Virginia ceded its claim to the territory north and west of the Ohio River, it never relinquished its claim on Kentucky, which officially remained part of Virginia until it was admitted as the fifteenth state of the Union in 1792. Jefferson being a Virginian, this effectively limited the area of his proposal to what was to become the Northwest Territory. Jefferson did extrapolate his proposed division all the way south to the Gulf Coast, but didn't name these other "states" in his final draft. This territory remained disputed between the United States, some states and several foreign powers (France, Spain, Great Britain). Later, the states of Kentucky, Tennessee, Alabama and Mississippi would be created here.

The month after he unveiled his plan, Jefferson left for France, for a five-year stint as ambassador. His proposal was superseded by the definitive Northwest Ordinance (1787) and forgotten. Maybe Jefferson took his precious draft with him to Paris. When an old French map of his proposed states was found in the New York State Library in 1856, no one had any idea whence the French had their strange idea.

As president (1801–9), Jefferson continued to further the westward expansion by the Louisiana Purchase (1803) and commissioned the Lewis and Clark expedition (1804).

3

Texas Can't Hold 'Em: The Lone Star Empire

Imperial Texas

Everything is bigger in Texas, the largest of the forty-eight contiguous states of the United States. But Texas itself could have been even bigger, as shown by this map. Texas had been settled by American colonists from the 1820s onward, and in 1836 declared its independence from Mexico. After defeating Mexican forces, the Republic of Texas had to choose between joining the United States or striking out on its own.

The former option was preferred by Texas's first president, Sam Houston (as by most Texans). But the United States hesitated, wary of war with Mexico and the prospect of adding yet another pro-slavery state. Texas withdrew its offer of annexation in 1838, when Houston was succeeded by Mirabeau Lamar, who favored the latter option.

Lamar obtained diplomatic recognition from several European countries and dreamed of an expanded Texas. Yet in the end, the pro-annexation camps won out in both the Texas and U.S. legislatures. Texas was annexed to the United States in 1845—although, intrigu-

ingly, the annexation may have been unconstitutional: The U.S. Congress approved it with a simple majority, but according to some legal experts should have done so with a two-thirds majority. But what if Texas had remained independent and prospered? Might it have looked like this?

The area marked "Republic of Texas" on this map was the core of the independent state, which also claimed territory farther west, disputed by Mexico, marked here as four counties: Presidio, El Paso, Worth and Santa Fe. The northern panhandle of the latter county reached all the way up to Mexico's then northernmost border.

After Texas's annexation, a large part of its unorganized west was transferred to the federal government, in exchange for taking on the republic's debt, reducing Texas to its present size. Today, Presidio and El Paso counties are part of the state of Texas, Worth County is entirely in New Mexico (as is part of the core area) and Santa Fe is scattered among Texas (its northern panhandle), New

Mexico, Oklahoma (its panhandle), Kansas, Colorado and a chunk of Wyoming.

If the Texan maximalists had had their way, Greater Texas would not only have remained independent, including the unorganized territories, but would also have laid claim to all the lands west thereof, including the Mexican Cession (i.e., the U.S. part of the claim) and large parts of northern Mexico, also to the south of Texas.

This would have created a veritable Lone Star Empire from the Pacific Ocean to the Gulf of Mexico. Maybe then the Texas embassy in London would still be there—nowadays the building opposite St. James's Palace houses a hat shop (although they might still stock some ten-gallon cowboy hats).

4

Western Designs on the Middle East: The Sykes-Picot Agreement

Palestinian Academic Society for the Study of International Affairs (PASSIA)

In November 1915, diplomats François Georges-Picot (for France) and Mark Sykes (for Britain) negotiated an "understanding" about how to divide the Middle East into spheres of influence for their respective countries. At the time, the area was still under control of the Ottoman Empire, linked to the Central Powers (Germany and Austria-Hungary) and therefore an opponent of the British, French and other Allies in World War I.

The Sykes-Picot Agreement was secretly consummated by the British and French governments on May 16, 1916. The outlines of the combined zones of influence have partially determined the borders of Syria, Israel, Jordan, Iraq and Saudi Arabia as they still stand today. Internally, the zones do not correspond to the present border situation.

According to Sykes-Picot, there were to be:

- A "Blue" zone of direct French control, in central Anatolia with extensions toward the south (the Syrian coast and Lebanon), the west (the southern Turkish coast) and far inland.
- A "Red" zone of direct British control, in southern Iraq, including Baghdad, and extending southward over Kuwait to include the Persian Gulf coast of Arabia.
- An "A" zone of French influence, somewhat corresponding with present-day Syria but without coastal access, and extending far into present-day Iraq, to include the city of Mosul.
- A "B" zone of British influence, roughly correspondent to present-day Jordan and Iraq and including the southern part of present-day Israel, down to the port of Haifa, and a band of territory extending deep into the Arabian Peninsula.
- An allied condominium in the Holy Land, pending consultation with other world powers.

France and Britain would be left free to decide on state boundaries within the areas of their control. The main criticism of the Sykes-Picot Agreement was that it failed to take into account the wishes of the Arab populations in the area—who had been promised self-determination by some Western interlocutors, such as Lawrence of Arabia, who promised the Arabs a homeland in exchange for siding with the British against the Turks.

The Sykes-Picot Agreement was later expanded to include Italy (which would receive some Aegean islands and a sphere of influence around Izmir/Smyrna on the Aegean coast of Asia Minor) and Russia (which would get Armenia and parts of Kurdistan). Due to the Communist Revolution of 1917, Russia's claims were denied. Italy's claims were formalized in the Treaty of Sèvres (1920).

Whether or not as "revenge," Lenin released a copy of the confidential agreement, causing great embarrassment among the Allies—and growing distrust among the Arabs. In fact, the Sykes-Picot Agreement is seen as a negative turning point in Arab-Western relations, which have not been "baggage-free" since.

Sykes-Picot was reaffirmed at the San Remo Conference (1920), although the borders of the resultant states (the French mandate area of Syria-Lebanon, the British mandate areas of Palestine, Transjordan, Iraq) did not correspond exactly to the zones of influence in the original agreement.

5

Franz Ferdinand's Winning Idea: The United States of Greater Austria

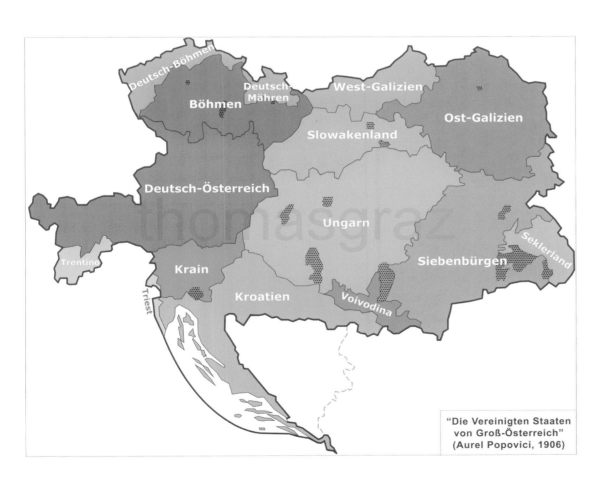

"Die Vereinigten Staaten
von Groß-Österreich"
(Aurel Popovici, 1906)

The First World War was triggered by the assassination of Franz Ferdinand in Sarajevo on June 28, 1914. Apart from the fact that a Scottish band was named after him, this is all most people know of the Austro-Hungarian archduke and crown prince. Here's another interesting piece of information about him:

Franz Ferdinand sponsored a group of scholars who came up with a plan to rearrange the volatile, multiethnic Double Monarchy into a United States of Greater Austria. The specific plan on this map was proposed in 1906 by Aurel Popovici, a Romanian subject of the empire. Had Franz Ferdinand acceded to the throne, would he have been willing and able to push through this kind of reform? And would a more equitably composed Austria-Hungary have been a viable state, maybe surviving until today?

The Austro-Hungarian Empire was unstable because it was dominated by only two out of its eleven constituent nationalities—Germans and Hungarians, together totaling only 44 percent of the empire's population. Resentment, resistance and revolt by the other nine nationalities made the situation untenable—and indeed, the empire imploded at the end of the war as swiftly as the Soviet Union in its final days.

To avoid ethnic strife, Popovici's plan proposed redrawing the internal boundaries of Austria-Hungary so that the provinces would be as ethnically and linguistically uniform as possible. Interspersing these homogenous states would be small autonomous "islands" where German-speakers dominated. Popovici defined the states as:

- German Austria (Deutsch-Österreich): present-day Austria, plus South Tyrol (now Italy) and neighboring Sudeten (i.e., German-speaking) areas of Bohemia (now the Czech Republic).
- German Bohemia (Deutsch-Böhmen): the northwestern part of the former Sudetenland, now the Czech Republic.
- German Moravia (Deutsch-Mähren): the northeastern part of the former Sudetenland, now in the Czech Republic.

- Bohemia (Böhmen): the Czech Republic of today, minus the formerly German-speaking areas specified above.
- Slovakia (Slowakenland): almost identical to the Slovak Republic of today.
- Western Galicia (West-Galizien): part of present-day Poland.
- Eastern Galicia (Ost-Galizien): part of present-day Ukraine.
- Hungary (Ungarn): slightly bigger than Hungary today, to the detriment of Slovakia and Serbia (Voivodina).
- Transylvania (Siebenbürgen): inhabited mainly by Romanians, but also by Hungarians and Germans.
- Székely Land (Seklerland): nowadays in central Romania, but still largely inhabited by the ethnic Hungarian Székely.
- Trento (Trentino): majority Italian-speaking.
- Trieste (Triest): autonomy was aimed at the Italian population, although the city also contained some Slovenians, Germans and Hungarians.
- Carniola (Krain): almost identical to Slovenia today.

- Croatia (Kroatien): almost identical to present-day Croatia.
- Voivodina: Serbo-Croatian for "Duchy"; an area with this name still is a multiethnic, autonomous region of Serbia, with a sizable Hungarian minority. Autonomy under Austria-Hungary would probably have been aimed primarily at placating the Serbian population.
- Autonomous regions for German speakers: would have been instituted in Bohemia, Slovakia, eastern Galicia and Carniola, with the most important of these areas in Hungary and Transylvania.

Because these proposed inner demarcations are along ethnic lines, they prefigure, however imperfectly, the national boundaries established in the region after the First World War. If you squint, you can already see the future borders, especially those of Czechoslovakia, Hungary, Austria, Yugoslavia and Romania.

6

Oklahoma's Stillborn Twin: The State of Sequoyah

When Oklahoma entered the Union in 1907, it could almost have done so as *two* states—Oklahoma and Sequoyah. The latter was a failed attempt to form a state for "Indians" in what is now eastern Oklahoma (where Native Americans still form a large part of the population).

Most of Oklahoma was set aside as Indian Territory by the Indian Removal Act of 1830, providing for the resettlement—voluntary and forced—of native tribes from east of the Mississippi. In 1866, the U.S. government seized the western half of the territory, which was gradually opened up for settlement from 1884 onward. In all, five "land runs" were organized. Some of the settlers were called "Sooners," because they had already literally staked their claim before the land was opened for settlement.

In 1890, the lands of the 1866 treaty, plus No Man's Land (now known as the Oklahoma Panhandle), were merged into the Oklahoma Territory. The eastern part of present-day Oklahoma remained Indian Territory. In a convention at Eufaula in 1902, representatives of the Five Civilized Tribes started a drive toward statehood for the Indian Territory. The name for their proposed state was Sequoyah, after a prominent Cherokee—in fact, the man who devised the Cherokee alphabet. In 1903, the delegates met again to organize a constitutional convention.

This convention met at Muskogee in 1905, presided over by General Pleasant Porter, principal chief of the Creek Nation. Vice presidents were the high representatives of each of the Five Civilized Tribes: William C. Rogers (Cherokee), William H. Murray (Chickasaw), Green McCurtain (Choctaw), John Brown (Seminole) and Charles N. Haskell (Creek). If Sequoyah never achieved statehood, it wasn't for the efforts of the convention: It drafted a constitution, established county boundaries for the new state, elected delegates to petition the U.S. Congress for statehood and saw its proposals overwhelmingly endorsed in a referendum held in Indian Territory.

However, eastern politicians pressured then U.S. president Theodore Roosevelt against admitting two western states (Sequoyah and Oklahoma) to the Union, fearing this would disproportionally diminish eastern states' political influence. Roosevelt then decided both territories could only enter the Union as a single state. Having already laid the groundwork for their own state, Indian Territory representatives had a big influence in establishing Oklahoma. The constitution of Oklahoma, admitted as the forth-sixth state in 1907, is based largely on that of Sequoyah.

The tantalizing concept of an "Indian" state of the Union was recycled by alternate history writer Harry Turtledove, in whose novel *How Few Remain* the Indian Territory enters the Confederate States of America as the Confederate State of Sequoyah.

Oklahoma today is an interesting blend of western and Native (or to use the less varnished vocabulary of yesteryear, "cowboy" and "Indian") cultures. Its name, first used by Chief Allen Wright of the Choctaw Nation during the 1866 treaty negotiations, means "Red People" in his native language. The state has the U.S.'s second largest Native American population, both percentage-wise (11.4 percent compared with Alaska's 19 percent) and in absolute terms (about 400,000, compared with California's 680,000). Oklahoma is home to about fifty Native tribal headquarters, more than any other state.

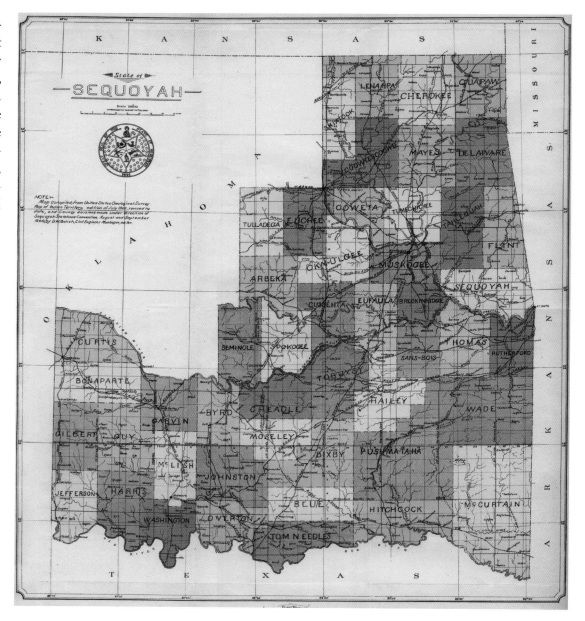

7

Dutch Dreams of Expansion into Germany: "Eastland, Our Land"

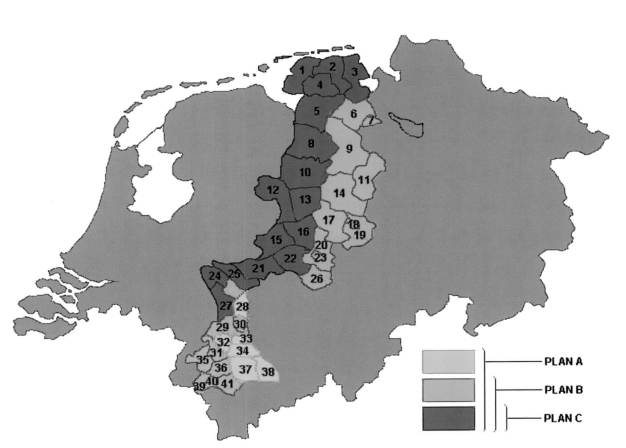

PLAN A
PLAN B
PLAN C

Immediately after World War II, there existed both official and unofficial plans in the Netherlands to annex a large area of Germany as a way of obtaining war reparations. A pivotal figure in most annexation schemes was Frits Bakker-Schut, a member of the State Commission for the Study of Annexation (Staatscommissie ter Bestudering van het Annexatievraagstuk, or SBA) and secretary of the (nongovernmental) Dutch Committee for Territorial Enlargement (Nederlandsch Comité voor Gebiedsuitbreiding, or NCG).

The SBA's target immediately after the war was to create momentum within the Netherlands for annexation of German territory. In brochures, it proposed the so-called Wezergrens (Weser Border, after the river). The slogan was, *Nederlands grens kome aan de Wezer* ("Let the Dutch Border Reach the Weser").

The NCG's task was to study in specific task forces the feasibility of the plan. The mineral wealth, agricultural possibilities and industrial potential for the intended areas were meticulously charted. The NCG presented its conclusion to the Dutch government at the end of 1945. It became known as the Bakker-Schut Plan, and proposed three formulas for annexation:

Plan A: annexation of all areas west of the line Wilhelmshaven-Osnabrück-Hamm-Wesel-Cologne-Aachen (including all those cities).

Plan B: basically the same proposal, but excluding the densely populated areas around Neuss, Mönchengladbach and Cologne from annexation.

Plan C: the smallest proposed area of annexation, with the border being moved to a line beginning in Varel, including all of Emsland and the Wesel area down toward Krefeld.

Apparently the plans included moves to "de-Germanize" the area, among other measures by giving towns a Dutch version of their German name. Some proposed place-name changes were (German name—Dutch name):

- Jülich—Gulik
- Emmerich—Emmerik
- Selfkant—Zelfkant
- Kleve—Kleef
- Aachen—Aken
- Bad Bentheim—Neder-Benthem
- Emlichheim—Emmelkamp
- Geilenkirchen—Geelkerken
- Geldern—Gelderen
- Goch—Gogh
- Moers—Meurs
- Münster—Munster
- Neuenhaus—Nieuwenhuis
- Nordhorn—Noordhoorn
- Osnabrück—Osnabrugge
- Veldhausen—Veldhuizen
- Wesel—Wezel
- Hoch-Elten—Hoog Elten
- Jemgum—Jemmingen
- Köln—Keulen
- Mönchen-Gladbach—Monniken-Glaabbeek
- Zwillbrock—Zwilbroek

Another measure to "Dutchify" the annexed area was to be population transfers (a bit like in the German areas to the east, which were annexed to Poland, Czechoslovakia and the Soviet Union). In the folder *Oostland—Ons Land* ("Eastland, Our Land"), the NCG proposed to expel all people from towns larger than 2,500 inhabitants, all former members of the Nazi Party and related organizations, and everybody who had settled in the area after 1933. The rest of the indigenous Germans would have the option of Dutch citizenship—if they spoke *plattdeutsch* (the local dialect, somewhat closer to Dutch than standard German) and had no close relatives in the rest of Germany. Everybody else was liable to be expelled without receiving compensation.

The Allied High Commission opposed the Dutch annexation plans on the grounds that Germany was already straining to accommo-date 14 million refugees from the east. More refugees from the west could destabilize further a situation urgently needing consolidation, to counter the growing Soviet threat on Western Europe. Interestingly, there was also a strong opposition to the plans within the Netherlands, particularly from the churches.

Nevertheless, at the Conference of the Western Occupying Powers of Germany in London (January 14–February 25, 1947), the Netherlands officially requested the annexation of 1,840 square kilometers (710 square miles) of German territory. This area, a modified and smaller version of the aforementioned Plan C, included the isle of Borkum, the county of Bentheim and a strip of border territory close to the cities of Ahaus, Rees, Kleve, Erkelenz, Geilenkirchen and Heinsberg. In 1946, the area housed about 160,000 people—over 90 percent German-speaking.

The concluding statements of the Germany Conference in London on April 23, 1949, awarded only very small fragments of German territory to the Netherlands—about twenty fragments, most smaller than one square kilometer and totaling no more than 69 square kilometers (26 square miles).

Most of these were returned to Germany in 1963 and 2002. The ambitious Dutch annexation plans of 1945 have resulted in only one formerly German area now still under Dutch control: a small area called Wylerberg (in German; Duivelsberg in Dutch) close to the Dutch border city of Nijmegen, measuring no more than 125 hectares (309 acres).

8

Sweet Home Talladego: C. Etzel Pearcy's Thirty-eight-State Union

The Proposed Thirty-Eight States of the United States of America

Even the most geographically challenged U.S. schoolkids—those who think Central America means Kansas—will be able to tell you how many states there are in the Union. Well they *should*, because fifty is a number so nice and big and round that it should be impossible to forget. For that reason alone, the proposal on this map seems not just unnecessary but foolish.

Why reduce the number of states from fifty to thirty-eight? Why replace every single one of the 104 existing state lines, all so familiar, many so pleasingly straight, with crooked borders that dissect all but six states? Why rename the remaining states without reusing a single of the newly extinct state names? This feels like wanton destruction.

This proposal was drawn up in 1973 by C. Etzel Pearcy, geography professor at California State University. Did some of that era's counterculture, counterintuition and counterintelligence drift in through his open office window? Did he share one doobie too many with his students? In fact, his proposal, which obviously never left the drawing board, makes more sense than a first glance warrants.

For a start, it makes practical sense. Pearcy contended that the existing state borders, many fixed before the land held significant numbers of people, now cut through large metropolitan areas, inviting administrative, political and economic problems. Why pay taxes for the infrastructure of the big city you work in if you live in the next state, for example?

Adding to this is that these all-too-straight borders also ignore more logical, natural boundaries such as rivers and mountains. Or maybe it's simply that straight and square were two very unpopular things to be in the 1970s, even for states. But Pearcy did have his eye on the money as well. Rationalizing the number of states would net the taxpayer a cool $4.6 billion, or $100 per citizen per year. And that's in 1973 money, mind you.

Pearcy attempted to draw his state lines through lesser-populated areas and mold each new state around one metropolis. He also eliminated "marginal protuberances" such as the Florida, Idaho, Maryland and Oklahoma panhandles. Pearcy's realignment also got rid of the small-size states in the Northeast, increasing the average state size by a quarter even while slicing Alaska in two and thus diminishing the extreme size difference of the present situation (Alaska is 483 times larger than Rhode Island).

If Professor Pearcy's proposal wasn't as potty as it initally appears, the absence of a powerful lobby for the implementation of his Thirty-eight State Plan in the intervening decades speaks volumes. People are just too attached to the names, histories and traditions of their states, take too much pride in their particularities and then are forced to work around the inconveniences caused by them. Or maybe they just prefer straight state lines to wobbly ones.

9

Forty-seven Fingers in the Antarctic Pie: The Frontage Principle

Seven countries currently claim territory in Antarctica. A map of these claims looks like a pie chart, as all are centered on the geographical South Pole. If those claims would be realized, you could visit six of those seven countries simply by walking in a small circle around that point (although "simple" is quite a euphemism for a stroll in one of the remotest, most hostile environments on the planet).

In what seems a very appropriate move, all claims on Antarctica are frozen by the Antarctic Treaty (1961), which also states that no future claims can be made. Present claimants are:

- United Kingdom (British Antarctic Territory, since 1908)
- New Zealand (Ross Dependency, since 1923)
- France (Terre Adélie, since 1924)
- Norway (Peter I Island, since 1929; Dronning Maud Land, since 1939)
- Australia (Australian Antarctic Territory, since 1933)

- Chile (Antártida Chilena, since 1940)
- Argentina (Antártida Argentina, since 1943)

All claims apply to the areas south of 60 degrees south, the northern limit of the Antarctic Treaty. The area between 60 degrees west and 150 west remains unclaimed, except for Peter I Island (Norway's claim to this territory is the only one in Antarctica that's not slice-shaped). Even though the Antarctic Treaty has managed to maintain the status quo since 1961, certain signatories—notably the United States and the Soviet Union (and its successor state Russia)—have stated that they do not feel bound by the interdiction of future claims.

A "claims race" could be very dangerous, as several of the already existing claims overlap, notably those of Chile, Argentina and Great Britain. The latter two already waged a brief war over the nearby (but non-Antarctic) Falkland Islands in 1982.

Terezinha de Castro, a Brazilian geostrategist, proposes a better method of divvying up the frozen South—a method that would eliminate the inherent danger of overlapping claims and, incidentally, would be more beneficial to Brazil's yet unrecognized Antarctic claims.

De Castro uses the principle of *defrontação*—frontage, in English. It works like this: The Antarctic coast between 0 degrees west to 90 degrees west (the South American sector) would be divided among South American nations according to the proportion of open-sea access their coastlines have to Antarctica. This would diminish the Argentine and Chilean sectors, give Uruguay, Peru and Ecuador a slice—and leave the biggest slice of frozen South Pole pie to Brazil.

An *Economist* article about the proposal gave Paul Youlten the idea to expand the frontage principle: "A useful little map got me wondering about which other countries might also lay claim to a slice of Antarctica, based on their having unrestricted southern passage to the continent across open seas."

The Youlten method of frontaging results in

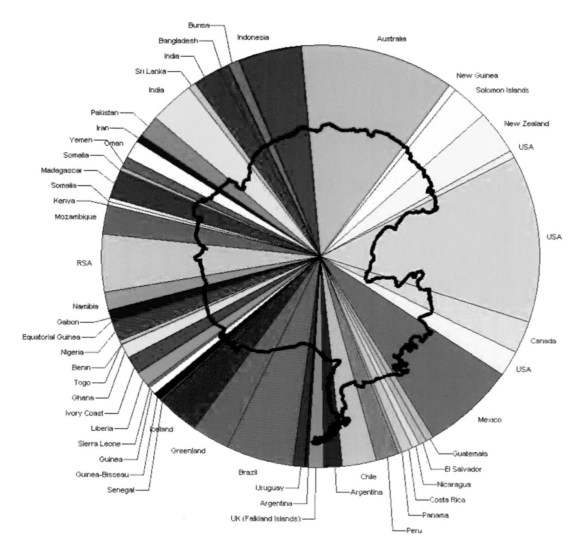

- Nine from Asia: Yemen, Oman, Iran, Pakistan, India, Sri Lanka, Bangladesh, Burma and Indonesia.
- Four from Oceania: Australia, Papua New Guinea, the Solomon Islands and New Zealand.
- Two from Europe: Iceland and the United Kingdom, albeit "indirectly," via the frontage of its South Atlantic possessions (the Falklands and South Georgia).

"Unexpected results include the surprising news that Somalia, Yemen and Oman could make claims," says Mr. Youlten. "As well as Iran, which I suspect might be tempted to set up an 'Icelamic Republic'—sorry, I couldn't resist that one."

Some countries get two sectors, because their frontage is split by another country's direct access to Antarctica (e.g., India's sectors, enveloping Sri Lanka's sector), or because they have noncontiguous territories, both with frontage (e.g., the United States, which has frontage through California and Alaska; the Alaskan sector is cut in two because Mr. Youlten used the 180 degrees meridian as his base line).

Youlten's map is not perfect, as he didn't include the frontage of smaller island nations such as Samoa, Cape Verde and the Maldives; the same goes for larger nations that possess islands with frontage, such as Portugal (the Azores), France (the Kerguelen Islands, New Caledonia) and Norway (Bouvet Island).

no fewer than forty-seven countries being able to claim their own slice of Antarctica. To wit:

- Fourteen from the Americas: Greenland, Canada, the United States, Mexico, Guatemala, El Salvador, Nicaragua, Costa Rica, Panama, Peru, Chile, Argentina, Uruguay and Brazil.

- Eighteen from Africa: Senegal, Guinea-Bissau, Guinea, Sierra Leone, Liberia, Côte d'Ivoire, Ghana, Togo, Benin, Nigeria, Equatorial Guinea, Gabon, Namibia, South Africa, Mozambique, Kenya, Somalia and Madagascar.

10

Out of One, Many: Sixteen New American Nations

The unofficial motto of the United States is *E Pluribus Unum*, "Out of Many, One." Matt Kirkland thinks the United States has become too unwieldy, and proposes to reverse the phrase—*Ex Unum, Pluribus* ("Out of One, Many").

Mr. Kirkland runs a Web site serving as "a grassroots movement, dedicated to breaking the United States into smaller, more functional nations." The site provides some extra information on each of the sixteen new, smaller American nations he proposes. But Mr. Kirkland doesn't have to have it all his own way. The site also contains "a fresh map, so that anyone can submit a new proposal."

His proposal comprises:

- Côte d'Atlantique (Maine): "When the New Nations are born, Cd'A plans to ally herself with Canada, eventually opting for voluntary annexation. Official language: French."
- New England (New Hampshire, Vermont, Massachusetts, Rhode Island, Connecticut and most of New York State): "New England expects to experience tense international relationships with its neighbors, New York, Jersey and Côte d'Atlantique."
- New York (New York City, Long Island, parts of New Jersey): "New Yorkers have neither the space nor the temperament for agriculture, and must import all foodstuffs."
- Jersey (Pennsylvania, Delaware, eastern Maryland, most of New Jersey): "Still smarting from losing Jersey City to the new nation of New York, Jerseyans plan to rebuild it—and call their capital New Jersey City."
- The Confederate States of the Atlantic (most of Virginia, North Carolina, South Carolina and Georgia): "The CSA is expected to adopt the Stars & Bars as a national flag at their first Confederation Conference."
- The Magic Kingdom of Florida (Florida): "Somewhat astonishingly, the Kingdom plans to squeeze the entire executive branch of government inside Cinderella's castle on the grounds of Walt Disney World."
- West Kendiano (Ohio, Indiana, Kentucky, Tennessee, West Virginia and the western part of Virginia): "While most citizens assume that their new name is an amalgamation of its components, West Kendiano actually refers to the now-extinct Kendiano Native Americans who originally occupied this territory."
- Soggy Bottom (Alabama, Mississippi, Louisiana): "Soggy Bottom will lead the New Nations among exporters of grits."
- The Boundary Waters (Michigan, Wisconsin, Minnesota): "Revolutionary sentiment in 'The Mitten' [i.e., southern Michigan], as its citizens prefer to call it, is growing. Only time will tell if the Boundary Waters can hold together as a nation."
- The People's Republic of the Plains (Illinois, Kansas, Arkansas, Oklahoma,

EX UNUM, PLURIBUS!

NORTHWEST TERRITORIES
pop. 12,339,798
★ Olympia

DAKOTA
pop. 1,395,173
★ Dakota City

THE BOUNDARY WATERS
pop. 20,511,362
★ St. Paul

COTE D'ATLANTIQUE (CANADA)
pop. 1,294,464
★ L'Amherst

NEW ENGLAND
pop. 20,007,209
★ Montpelier

NEW YORK
pop. 12,231,564
★ New York City

CALIVADA
pop. 37,289,524
★ Sacramento

FOUR CORNERS
pop. 14,134,310
★ Four Corners

THE PEOPLES REPUBLIC OF THE PLAINS
pop. 31,858,816
★ Chicago
★ Branson

WEST KENDIANO
pop. 29,272,388
★ Cincinnati

JERSEY
pop. 22,303,674
★ New Jersey City

THE CONFEDERATE STATES OF THE ATLANTIC
pop. 33,739,318
★ Columbia

EL REPUBLICO DE TEJAS
pop. 20,511,362
★ Waco

SOGGY BOTTOM
pop. 11,840,936
★ New Orleans

THE MAGIC KINGDOM OF FLORIDA
pop. 16,713,149
★ Walt Disney World

HA'AWASKA
pop. 1,888,684
★ Honolulu
★ Anchorage

Nebraska, Missouri, Iowa): "The PRP expects to dominate the annual International American Football Association championship tournament."

- El Republico de Tejas (Texas): "Tejanos originally fought the proposals to dissolve the US, arguing they were never really part of the Union anyway."
- Dakota (North and South Dakota): "With their share of the spoils of the defunct federal government, Dakotans plan to build a shining example of a well-planned capital. Dakota City will host 85% of the national population."

  Another fun fact: "Dakotans have proposed a revolutionary new system for their currency. Paper denominations of the 'dakot' will be numbered according to the primes and coins—one hundred 'iotas' equal a 'dakot'—will follow the fibonacci sequence. Math skills are expected to skyrocket as a result."
- Northwest Territories (Washington, Oregon, Idaho, Montana and Wyoming): "Only in theory will Olympia's

governmental powers reach past the Sierra Nevada. Most of the eastern high plains will most likely be controlled (peaceably) by independent militias."

- Calivada (California and Nevada): "After the dissolution of the US, Calivada will hold title to the world's second largest economy." (Pop.: 37.3 million)
- Four Corners (Utah, Colorado, Arizona and New Mexico): "Once construction is completed, the Parlia-Dome of the Four Corners Capitol building will sit exactly at the juncture of its component states. Members will be able to sit through an entire session of parliament without actually leaving their state's territory."
- Ha'awaska (Hawaii and Alaska): "Ha'awaska employs a bicameral capital system, keeping governmental functions cordoned off in Honolulu and Anchorage."

Some readers pointed out factual errors on the map, (in correct Spanish, it would be "La republica de Tejas"; Montpelier seems to be

misplaced in New Hampshire; and the French themselves would write "Côte atlantique").

Others challenge the proposed divisions. Why would New York, even upstate, feel inclined to join the historically distinct region of New England? Wouldn't it be better to lump the Dakotas, Montana, Idaho and Wyoming together, and group Washington, Oregon and Northern California as "Cascadia" or "Ecotopia"? Why do Ha'awaska and the PRP need two capitals each? Shouldn't "Columbus" be "Columbia," capital of South Carolina? Do you really think anyone in Pennsylvania would want to belong to something called "Jersey"?

And finally, my favorite: If Cincinnati is so easy to misspell, maybe it should revert to its original name: Losantiville.

Joel Garreau's famous 1970s map of the Nine Nations of North America may have been economically and culturally better grounded. This map is only one of many that prove that the tension between the U.S.'s institutional unity and its practical diversity sometimes leads to very strange ideas indeed.

11

The Fro Gymraeg: A Reservation for Welsh

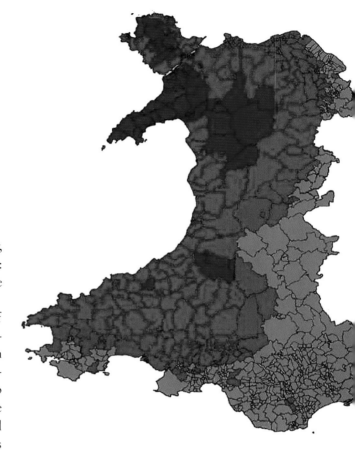

English is the dominant language in the British Isles, even in the so-called Celtic fringe—Ireland, Scotland and Wales. In Scotland but mainly in Ireland, some territorial measures have been taken to protect the indigenous language, especially in the areas where it remains strongest.

Those areas in Ireland are called the Gaeltacht, a collection of noncontiguous rural and mainly western language islands where Irish Gaelic is the official first language. In Scotland, a corresponding term, Gàidhealtachd, is used to describe the area in the northern Highlands where Celtic culture is strongest—Scottish Gaelic even there being almost extinct.

In Wales, which for centuries formed one legal entity with England (but for a few years elects its own Welsh assembly, which has limited powers), no such "language reservation" has been designated yet. An advocacy group called Cymuned (Welsh for "community") campaigns for setting up an area similar to the Irish Gaeltacht. As with the Celtic languages in Ireland and Scotland, Welsh is under con-

stant assault from English, not just culturally, but also in an economic/demographic way: Many English move into Wales, lured by the lower cost of housing.

Cymuned advocates the establishment of an area to be called Fro Gymraeg, in which special provisions for the survival and promotion of the Welsh language (Cymraeg) can be implemented. The Fro Gymraeg is to be made up of those areas where at least 50 percent of the natives speak Welsh. Those areas are marked light red on this map (dark red indicates areas where at least 80 percent speak Welsh).

In the darker green areas, over 20 percent of Welsh-born people speak Welsh, and "support should be made available for them to work towards becoming part of the Fro if that is what they desire," this Web site states. Inside the Fro Gymraeg, Cymuned would like to implement, among others, these measures:

- An elected Statutory Council to represent the Fro.
- Planning permission for local people only.

- Cymraeg to be the internal language of local government.
- Individuals who provide statutory public services should speak Welsh.
- Welsh history and language citizenship lessons should be available for incomers.
- Cymraeg should be the medium of education for all students between three and sixteen.
- To aim at extending the Fro Gymraeg through helping electoral wards outside the Fro to vote to become a part of the Fro.

VIII. EPHEMERAL STATES

*These countries did at some point exist, but so fleetingly
that they have now been all but forgotten.*

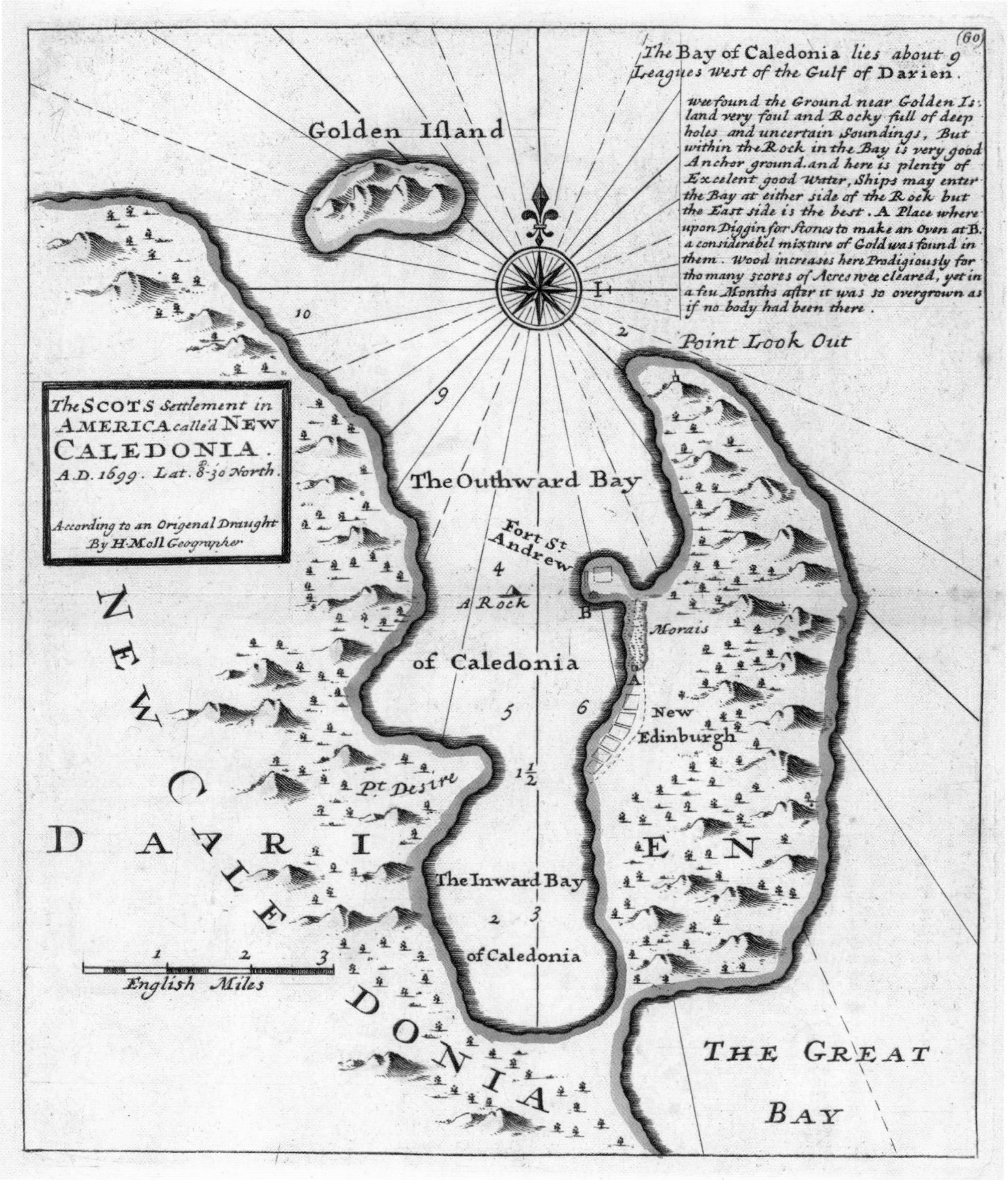

The Bay of Caledonia lies about 9 Leagues west of the Gulf of Darien.

wee found the Ground near Golden Island very foul and Rocky full of deep holes and uncertain soundings, But within the Rock in the Bay is very good Anchor ground. and here is plenty of Excelent good Water, Ships may enter the Bay at either side of the Rock but the East side is the best. A Place where upon Diggin for stones to make an Oven at B. a considerabel mixture of Gold was found in them. Wood increases here Prodigiously for tho many scores of Acres wee cleared, yet in a feu Months after it was so overgrown as if no body had been there.

Golden Island

The SCOTS Settlement in AMERICA call'd NEW CALEDONIA. A.D. 1699. Lat. 8.30 North.

According to an Origenal Draught By H. Moll Geographer.

Point Look Out

The Outhward Bay

Fort St Andrew

A Rock

B

of Caledonia

Morais

A

New Edinburgh

NEW

CALE

DARI

EN

Pt Desire

The Inward Bay

of Caledonia

DONIA

THE GREAT

BAY

English Miles

1 2 3

10 9 5 6 1½ 4 2 2 3

1

The Darién Scheme: A Tropical Scotland Gone Horribly Wrong

Scotland's last, worst project as an independent nation was an attempt at establishing a colony in Central America. This Darién Scheme, so called after the hot, humid and disease-ridden isthmus in present-day Panama, failed so spectacularly that it allowed England to force the Act of Union upon its northern neighbor.

Ever since the Union of the Crowns (1603), Scotland and England had been ruled by the same monarchs, but each country still had its own parliament—and nominal independence. This meant, among other things, that Scotland was excluded from the lucrative New World commerce, jealously guarded by the East India Company (EIC) and other English trading associations.

At the end of the seventeenth century, a stalling economy and seven successive years of failed harvests whetted Scotland's appetite for a foreign adventure of its own. In 1685, the Scottish parliament granted the newly formed "Company of Scotland Trading to Africa and the Indies" a thirty-one-year monopoly on trade with the colonies. All that was needed now were . . . colonies.

Scotland's only other colony had been Nova Scotia, lost to France in 1632 (by the *English*, mind you). This time around, the plan was to set up a new Scotland by another name (i.e., New Caledonia) as a trading colony on the Darién isthmus. This scheme was the brainchild of William Paterson, a Scotsman who had previously set up the Bank of England (in 1694).

Paterson, whose naïveté was only exceeded by his enthusiasm, declared Darién "the door of the seas, the key of the universe"—without ever having been there. In fact, the same isthmus would two centuries later be sliced by the Panama Canal, and for the same reason: to connect trade from the Atlantic and Pacific oceans. The Scots intended to do this via overland caravans.

The Darién Scheme was either a brilliant strategic move made centuries too soon, or a completely stupid plan at exactly the right time. The Scottish public, however, was convinced it was brilliant *and* timely: Literally half the population of Lowland Scotland rushed to invest in it. Initial English investors were pressured to withdraw from the scheme by the English parliament, which feared competition for the EIC.

Scotland pressed on regardless (also choosing to ignore the Spanish claim to the intended Scottish colony), and in July 1698, a first expedition of five ships sailed forth from Leith, carrying twelve hundred colonists—none of whom knew their final destination. Only past Madeira did the ships' captains break the seals of their secret orders, directing them to the Bay of Darién and the Golden Island, "to the leeward mouth of the River of Darién, and there make a settlement on the mainland."

The colonists reached New Caledonia at the end of October, built a fort they named St. Andrew (after Scotland's patron saint) and a handful of palmetto huts they ambitiously called New Edinburgh. The first progress reports were glowingly positive: The land was fertile, the fruits exotic and the opportunities for hunting and fishing plentiful.

A dramatic exaggeration, as it soon turned out. Agriculture in the damp, oppressive climate proved nigh on impossible, the local Indians weren't interested in trading for trinkets, and worst of all, there were no ships to trade with. Paterson's wife died a few days after landfall, sending him on a downward spiral of depression. King William forbade the English colonies in the Americas to offer material help to the Scottish colony.

Heat, humidity and hunger soon killed four hundred colonists. Those who weren't dead suffered from famine and yellow fever. To make matters worse, the Spanish had discovered and attacked New Caledonia. In mid-1699, the survivors gave up and sought refuge in English-American colonies, from Jamaica to New York.

News—good and bad—traveled slowly in those days, so in August 1699 a second group of thirteen hundred colonists left Scotland, soon followed by a third one. This second wave found four hundred fresh graves, and the ruins of New Edinburgh already being overwhelmed by the "vast, howling wilderness." And yet they decided to give New Caledonia another try. They even preemptively attacked a Spanish fort—a successful strike that marks the only military victory of this Scottish colonial adventure. For after a monthlong Spanish siege of Fort St. Andrew, the Scots gave up and went home. New Caledonia was abandoned for the last time in April 1700, again to be overwhelmed by the inhospitable jungle.

The disaster was total. Out of sixteen ships that had been sent to Darién, only one made it back to Scotland. Over two thousand colonists never made it home: Many had died, others were forced into near-slavery on plantations. Scottish investors had lost over £400,000—half the liquid capital available in Scotland at the time.

The failure of the Darién Scheme had two opposite effects. First, it caused a wave of anti-English sentiment, as England was blamed for actively obstructing Scotland's attempt at colonization. One English captain was even hanged on trumped-up charges of piracy. Second, the king proposed an Act of Union that explicitly promised to pay back the lost investment, plus interest.

This instance of overt, institutional bribery turned many disappointed Scottish investors into fervent Unionists. In 1707, the Act of Union was adopted by both parliaments, abolishing the Scottish parliament (Scottish MPs would have to travel to Westminster from now on) and establishing the United Kingdom.

Only in 1999 was a separate Scottish parliament reinstated. New Caledonia resurfaced half a world away, when in 1774 Captain Cook gave the name to an island group in the Pacific. La Nouvelle Calédonie is now a French overseas territory.

2

Gray Area Between the United States and Canada: The Republic of Indian Stream (1832–35)

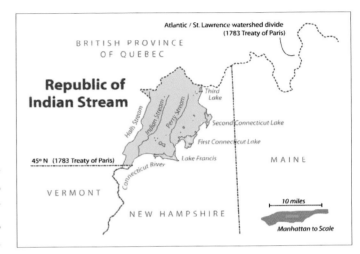

The Republic of Indian Stream (RoIS) was a self-declared (but unrecognized) republic in a "gray area" between the United States and Canada from 1832 to 1835. Although very small and sparsely populated (the "Streamers" never numbered more than about three hundred), the RoIS boasted a constitution and an elected government.

The "gray area" in which the RoIS was established resulted from an ambiguity in the Treaty of Paris (1783), which established the border between the newly independent United States and the remainder of British North America. It defined the border between the United States and (what was to become) Canada in the north of New Hampshire as *the northwesternmost head of the Connecticut River*. There are, however, several possibilities, as shown on the map: the heads of Hall's, Indian and Perry streams, and Third Lake, the origin of the Connecticut River itself.

Obviously, the United States and Britain each interpreted the ambiguity to their maxi-

mum advantage, the United States considering Hall's Stream the border between the two states, and Britain opting for the waterway beginning at Third Lake. As a result of this, the area in between was neither here nor there. Except for tax purposes. *Both* states sent tax and debt collectors into the area—which chagrined the inhabitants so much that they declared their independence . . . but only until the Americans and the British could sort out their differences.

Things came to a head when a band of Streamers invaded Canada to liberate one of their countrymen from custody. This particular Streamer had been arrested by a British sheriff because of an unpaid hardware store debt. The invading posse shot up the judge's house where their compatriot was held. This caused an international incident—even though the idea of a war caused by an unpaid store debt did seem a bit ludicrous.

As the Brits and Yanks agreed to resolve this particular border dispute, the Streamers

hastily voted to be annexed by the United States The New Hampshire Militia occupied the area shortly thereafter. Britain relinquished its claim in 1836 and the border was established according to the "maximalist" American interpretation of the Treaty of Paris.

In 1840, the area of the RoIS was incorporated as the township of Pittsburg, which still is the largest township in the United States, covering about 750 square kilometers (290 square miles). The dispute was definitively resolved in 1842 in the Webster-Ashburton Treaty, which mainly dealt with the establishment of the boundary between (what were to become) the U.S. state of Maine and the Canadian province of New Brunswick.

3

A Small Country That Never Was: Greater Belgium

This map of the Low Countries was outdated by the time it was published in 1840. And even in the preceding years, it never corresponded to the situation on the ground: Belgium never was this big.

Although the country's name has a long pedigree—Julius Caesar mentions the Belgae as the "bravest of all Gauls"—as European nations go, Belgium is one of the young'uns. It achieved independence only in 1830 and, unique among nations, as a result of an opera. At a performance of Daniel Auber's grand opera *La muette de Portici* (The Mute Woman of Portici) in Brussels on August 25, 1830, the audience identified with the Neapolitan fight against Spanish tyranny. The ensuing anti-Dutch riot quickly escalated into a full-blown revolution, and eventually led to Belgium's independence from the Netherlands.

In the Middle Ages, Belgium, the Netherlands and Luxembourg had formed a territorial unit under the dukes of Burgundy. But the religious wars of the sixteenth century led to

an independent, Protestant, prosperous north and a south that would remain Catholic—and under foreign domination until 1830.

After centuries of Spanish, Austrian and French overlordship, Dutch rule was imposed on the Southern Netherlands in 1815, after Napoleon's defeat at Waterloo. This United Kingdom of the Netherlands was to be a strong buffer state against future French aggression, but the centuries of separation proved impossible to surmount. The north was unilingually Dutch-speaking and mainly Protestant, and the south prominently Francophone and almost exclusively Catholic.

Dutch maladroitness and massive desertions from the army (two-thirds of which was made up of "southerners") fueled the operatically inspired revolution, which by the end of the year had managed to "liberate" the entire area colored yellow on this map—with a few notable exceptions: the citadel of Antwerp, the cities of Maastricht and Luxembourg, and the commune of Mook.

In August 1831, the north counterattacked.

In a campaign lasting a mere ten days, King William managed to defeat the Belgian rebels several times. But France, which for obvious reasons supported Belgian independence, sent its own army across the border. William then quickly concluded an armistice with the Belgians and retreated north.

Nevertheless, the Dutch victory proved to be more than pyrrhic—the terms of separation were altered to the north's advantage. Dutch garrisons were posted indefinitely in Maastricht and Luxembourg. This situation would be consolidated and expanded to the detriment of Belgium at the Treaty of London (1839). By this agreement, Belgium's independence was recognized and guaranteed by the United Kingdom, France, Prussia, Austria and Russia—and, crucially, the Netherlands.

It replaced an earlier treaty, refused by the Dutch because it was too favorable for the Belgians. This new treaty allowed the Dutch to reacquire parts of two southern provinces—not coincidentally around both towns where the Dutch had been able to maintain garrisons. Luxembourg was split into a western, Belgian part and an "independent" eastern part (which would remain joined to the Netherlands in a personal union with its monarch). The split was largely identical to the French-German language border splitting the province.

Limburg was divided into a Belgian east and a Dutch west, the border being formed by the Meuse—except where Maastricht straddles the river. The distance between the city center and the border was established as a bit more than the range of contemporary cannon. All of which explains:

- Why Luxembourg still shares a flag with the Netherlands (the personal union ended in 1890, when the Dutch king William III died childless);
- Why there are two sets of Limburgs and Luxembourgs in and near Belgium;
- Why Belgium, compared with this map, misses two bits of territory in the east (coloration on this map was a bit sloppy; the northernmost point Belgium would have achieved is the aforementioned commune of Mook en Middelaar, just south of Nijmegen on the German-Dutch border).

All of which does not explain:

- Why Belgium seems to miss another piece of territory in the east. Only after the First World War did Belgium achieve its present territory, when it annexed a few German-speaking districts.
- Why Belgium didn't have any claims

on the area south of the Scheldt River, potentially obstructing shipping to and from Antwerp. In contrast to most of Limburg, which was Catholic and had a "southern" mentality, this part of Zeeland was—and still is—staunchly Protestant, and largely unsympathetic to the south's cause.

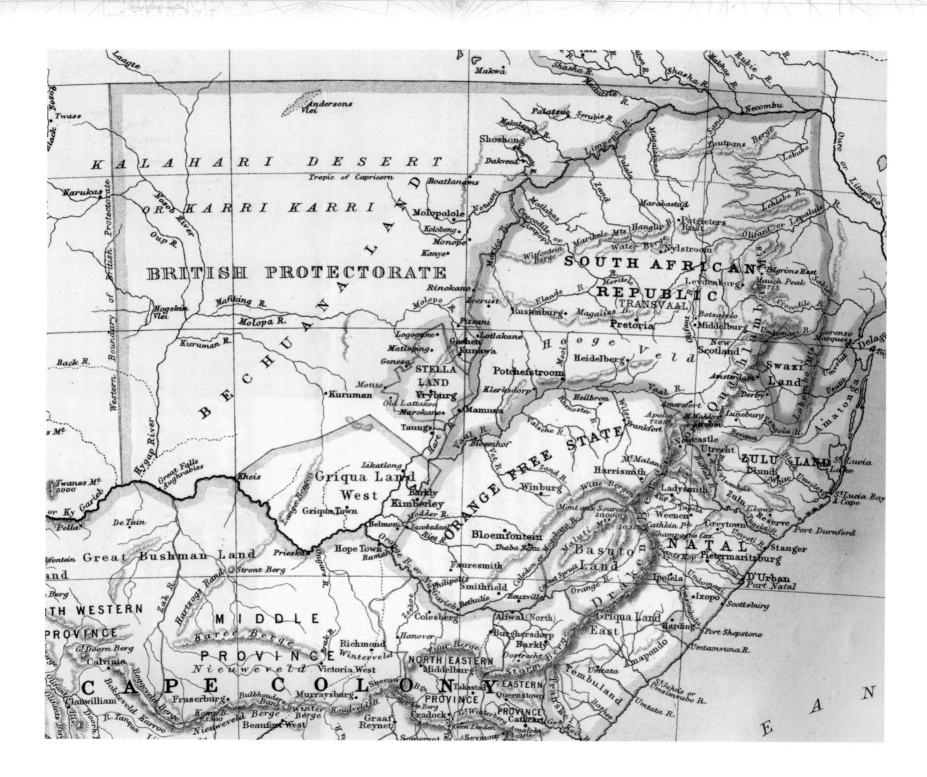

KALAHARI DESERT

KARRI KARRI

BRITISH PROTECTORATE

SOUTH AFRICAN
REPUBLIC
(TRANSVAAL)

Pretoria

STELLA
LAND
Vryburg

Griqua Land
West

Kimberley

ORANGE FREE STATE

Bloemfontein

Swazi
Land

ZULU LAND

NATAL

Basuto
Land

Pietermaritzburg

D'Urban
Port Natal

Great Bushman Land

MIDDLE

PROVINCE

CAPE COLONY

NORTH EASTERN
PROVINCE

EASTERN
PROVINCE

Griqua Land
East

Amapondo

4

Nice Name for It: The United States of Stellaland

Nowadays, the southern tip of Africa is dominated by a single state, the Republic of South Africa (punctuated by Lesotho, one of the world's few enclave states). But starting about a century and a half ago, when the usurping British were pushing the Dutch-originated Afrikaners inland, the eastern part of the RSA's present territory was littered with a number of *Boererepublieke* ("boer" means "farmer," but became synonymous with white, Afrikaans-speaking and anti-British).

These republics were later annexed by the British, after two Anglo-Boer Wars at the end of the nineteenth and the beginning of the twentieth century. The largest and best known of these republics became constituent provinces of the Union of South Africa (later the Republic of South Africa): the Orange Free State, Transvaal and Natal. But there were also smaller *Boererepublieke* that just disappeared off the map, including the intriguingly small and short-lived United States of Stellaland (1882–85).

Stellaland owes its existence to the war between the Batlaping and the Korannas, black tribes that had both hired white mercenaries. David Massouw, leader of the Korannas, had promised the Boers homesteads if they helped him win the war. After the war ended in July 1882, these homesteads were granted to exactly 416 white farmers, who thereafter considered themselves "free citizens" and formed the independent republic of Stellaland on July 26, 1882.

The name was chosen to refer to the comet that was visible in the sky at the time of the decisive battle (although Stella is Latin for "star," not for "comet"). The capital city was called Vryburg (Freetown), on a place known to the Tswana as Huhudi (Running Water). First and only president was Gerrit Jacobus van Niekerk (1849–1896). Stellaland expanded to include the neighboring boer republic of Goosen. The two nations were known collectively as the United States of Stellaland.

Stellaland aspired to be united with the big boer republic to the east, Transvaal. The British government, then in control of the formerly Dutch Cape Province, objected to the westwardly expansion of Transvaal and decided to invade. An expeditionary force under Sir Charles Warren entered the territory in February 1885, and it was formally annexed to British Bechuanaland on September 30, 1885.

During apartheid, the area around Vryburg was a "white" island in the (nominally) independent Bantustan of Bophutatswana. Since 1994, when the RSA's administrative divisions were reorganized following the end of apartheid, the area is part of the North West Province of the RSA. Today, the name Stellaland is still used to refer to the area around the villages of Vryburg, Stella and Reivilo.

Stellaland has had three different flags in its short existence, the first being the state emblem on a green background, the second a six-pointed white star on the same green and the third an eight-pointed white star on a field split vertically between green (left) and red (right). One of the reasons for this diversity is that apparently the president's wife had to make all those flags herself, and didn't always have the right material to copy a previous design.

5

A Pizza Slice Called Friendship: The World's First Esperanto State

The story of Amikejo is a fantastic piece of obscure cartographic and cultural history: Amikejo was the world's first and only state based on the ideals of the Esperantist movement. It was founded in a tiny (1.35 square miles; 3.5 sq. km), pizza-slice-shaped area that for a hundred years was an easily overlooked "neutral zone" in western Europe.

To find the general area where this neutral zone once was, take a map of Europe and find the point where the Netherlands, Germany and Belgium meet. This *Drielandenpunt* ("trinational point" in Dutch) even today is a bizarre enough place in itself.

It's the southernmost point of the Netherlands, a country world-renowned for its flatness, and is at the same time its *highest* point. The German city of Aachen, once the capital of Charlemagne's empire, is a mere three miles away. And across the Belgian border lies a hazy zone of transition between Germanic and Latin cultures, *and* Dutch, French and German language zones.

This is where it gets really weird: This *Drielandenpunt* once was a *Vierlandenpunt* ("quadrinational point")—arguably the only one in the world ever. How did this come about?

After Napoleon's defeat at Waterloo in 1814, the Congress of Vienna redrew the map of Europe. The United Kingdom of the Netherlands was constituted as an anti-French buffer state, consisting roughly of the three present-day Benelux states (Belgium, the Netherlands and Luxembourg). Its border with the German state of Prussia was left undefined in the area of Moresnet, because of an important zinc mine claimed by both powers.

In a separate treaty, the Dutch and the Prussians divided Moresnet into a Dutch, a Prussian and a neutral zone. The latter was to be administered by two commissars from each country. The boundary point at Vaals became a "trinational" point between Prussia, the Netherlands and Neutral Moresnet.

When Belgium seceded from the United Kingdom of the Netherlands in 1830, it assumed control over "Dutch" Moresnet and gained the commissary rights to Neutral Moresnet. The trinational point at Vaals became a quadrinational point.

Due to the economic good fortunes of Vieille Montagne, the local zinc mine, the number of inhabitants of Neutral Moresnet grew fivefold from 500 (in 1850) to more than 2,500 in 1856. Living in neutral territory had pluses and minuses. These "neutrals" could escape military service in the surrounding countries, for example, but were stateless when they traveled "abroad."

Wilhelm Moly, a German doctor, and Gustave Roy, a French professor—both keen Esperantists—decided in 1906 to establish an Esperanto state in Neutral Moresnet. Esperanto, the artificial language developed some decades before by L. L. Zamenhof, a Polish doctor, was supposed to transcend the linguistic divides crippling Europe.

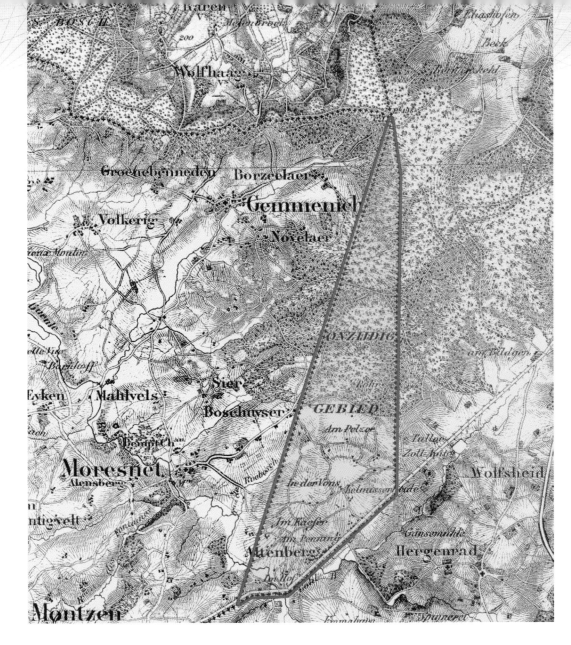

In 1908, a great demonstration was held in Neutral Moresnet, attended by the whole population, advocating the establishment of an Esperanto Free State to be called Amikejo (Esperanto for "Place of Friendship"). The local band played a tune that would be the national anthem of Amikejo.

In the meantime, tensions had been building between Belgium and Prussia/Germany over the neutral territory (which had outlasted its usefulness since the depletion of the zinc mine). The locals petitioned Belgium for annexation, following some strong-arm tactics by Prussia/Germany.

In 1919, the Treaty of Versailles officialized the annexation of the territory to Belgium, thus ending its neutral state. It's unclear what happened to Amikejo, although it's likely its high-minded idealism was simply swept away by the brutal forces of war.

6

Independent for Only Twenty-four Hours: Carpatho-Ukraine

Carpatho-Ukraine must have been the shortest-lived state in history. It existed for no more than twenty-four hours, declaring its independence from Czechoslovakia on March 15, 1939, and being formally annexed by Hungary one day later.

Predominantly inhabited by Ruthenians (Ukrainians), Subcarpathian Ruthenia (or Transcarpathia) came to be the very tail end of the snakelike postwar construction known as Czechoslovakia after the collapse of the Austro-Hungarian Empire in 1918.

When in 1938 Nazi Germany annexed the Sudetenland parts in the west of the country, thus weakening the integrity of the Czechoslovak state, Transcarpathia (and Slovakia) demanded and got more autonomy. The region renamed itself Carpathian Ruthenia, and in November 1938 changed its name again to Carpatho-Ukraine.

In that same month, its southern part (and the southern third of Slovakia) was annexed by Hungary, an ally of Nazi Germany. This did not calm the Czechoslovak-Hungarian tensions: between November 2, 1938, and January 12, 1939, twenty-two border clashes ensued. These clashes, and the ineffectual response of the Czech-dominated state, further encouraged separatism in the Slovak and Ruthenian east of Czechoslovakia.

Not satisfied with mere autonomy and encouraged by the Nazis, the Slovaks declared full independence on March 14, 1939. The next day, Hitler had his troops march into Bohemia and Moravia, the Czech part of Czechoslovakia.

This left Carpatho-Ukraine no other option than to declare its own independence, which occurred on March 15, 1939. The first (and only) president of the Republic of Carpatho-Ukraine was the Reverend Avhustyn Voloshyn. The declaration of independence immediately plunged the new state into anarchy, as irregular troops staged terrorist attacks against the remnants of the Czech army, as well as against pro-Slovak and pro-Hungarian parts of the population.

This, and further border skirmishes, proved sufficient cause for Hungary to invade the region. This happened on the same day as the declaration of independence. One day later, Hungary formally annexed the whole territory. On March 17, the Hungarian troops reached the Polish border. The last resistance was broken the next day.

In 1944, advancing Soviet troops refused Czechoslovak government officials permission to resume control over the area. In June 1945, a treaty between Czechoslovakia and the Soviet Union stipulated that the area was to become part of the Ukrainian Soviet Socialist Republic, under again another name: the Zakarpathia Oblast.

IX. STRANGE BORDERS

Borders are often mistaken for mundane instruments of demarcation,
when in fact they are trampolines for the imagination.

1

The Circle and the Wedge: Delaware's Curious Border

The northern border of Delaware is characterized by two geopolitical anomalies:

- The Twelve-Mile Circle, the border with Pennsylvania and the only circular boundary between U.S. states. It is centered on the cupola of the New Castle courthouse and dates back to the original deed of Delaware by the Duke of York to William Penn.
- The Delaware Wedge, an area of about one square mile (3 sq. km) "plugging" the gap between the western part of the Circle and the northeastern edge of the Maryland border. The Wedge was disputed territory until 1921, when Pennsylvania recognized Delaware's claim to it.

The origins of the Circle and the Wedge shed light on the importance of (or lack of) exact surveying methods in establishing colonial boundaries.

The Maryland Charter of 1632 gave the Calvert family, already owners of Maryland, possession of the entire Delmarva Peninsula up to the 40th parallel. However, in 1664, the Duke of York decided to separate the area around New Castle and lands to the south thereof from Maryland, to be administered as a new colony.

The Pennsylvania Charter of 1681 gave William Penn possession of land west of the Delaware River and north of the 40th parallel, excluding any land in a twelve-mile radius around New Castle (the town is actually twenty-five miles south of the 40th parallel, indicating how inaccurately the land was surveyed). The Penn family later acquired New Castle and adjoining territories, but would continue to administer these "three lower counties" as a separate colony.

The borders between Maryland, Pennsylvania and the "three lower counties" remained vague. In fact, only after Philadelphia was settled did the Penns discover it was south

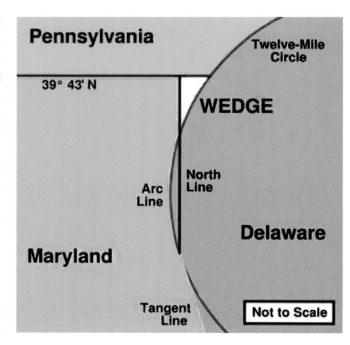

of the 40th parallel and tried to extend their claim accordingly. In the 1730s, the border dispute would lead to armed conflict between the two colonies, known as Cresap's War.

Surveys established the Transpeninsular Line as Delaware's southern border, and the Twelve-Mile Circle as its northern border. From these two main lines, the Penns and Calverts derived other border lines:

- a Tangent Line, connecting the middle of the Transpeninsular Line with the western side of the Twelve-Mile Circle, separating Delaware from Maryland; and
- a North Line, connecting the point where Tangent meets Circle with a parallel line at 39°43' N (fifteen miles south of Philadelphia, as a compromise to the 40th parallel).

After initially refusing to recognize the border, Maryland was forced to do so by the king, after the border was resurveyed by Mason and Dixon—establishing the Tangent Line and the Compromise Parallel as the "original" Mason-Dixon Line (which has since become the proverbial border between North and South).

It was agreed that any part inside the Arc Line west of the North Line would remain part of Delaware. The surveyors hadn't counted on the area *east* of the North Line but *outside* the Arc Line and *below* the compromise parallel.

This wedge-shaped gray area was not an issue for as long as the Penns owned both Pennsylvania and Delaware. It became a hot potato when they became separate states. Pennsylvania claimed the Wedge because it lay beyond the Twelve-Mile Circle; Delaware did the same because the area lay south of 39°43' N latitude.

Maryland did not have any claim, because its eastern border was defined by the North Line (and the Arc Line protruding from it). Only in 1921 did Pennsylvania relent and let Delaware have undisputed possession of the Wedge and the town of Mechanicsville, which lies within the area.

2

Going, Going, Gone: The Old Cherokee Country

The Cherokee, with 250,000 members of federally recognized Nations and Bands (2000 census), are among the most numerous of all remaining Native American (i.e., "Indian") tribes in the United States. At the time of European contact in the sixteenth century, they lived mainly in what is now the southeastern United States. Contact often took the shape of conflict, and by the early eighteenth century, a united Cherokee empire under Moytoy I and II frequently was at odds with European settlers in the Carolinas.

The latter emperor's acceptance of King George II of England's suzerainty, around 1730, didn't stop European encroachment on traditional Cherokee lands. Some Cherokee headed out west, far away from the colonists, establishing settlements across the Mississippi in what is now Arkansas. But by the 1820s, these Cherokee were again pushed farther west by white settlers, to what is now Oklahoma.

In the 1830s, a gold rush in Georgia drew thousands of Europeans into the Cherokee heartland. Despite a Supreme Court ruling in their favor, the remaining Cherokee were forc-ibly expatriated and sent westward, on what became known as the Trail of Tears. Many perished on the road, and not being able to give these dead their traditional burial, the singing of "Amazing Grace" had to suffice. This song has been considered the Cherokee national anthem ever since.

Some Cherokee managed to escape deportation via the Trail of Tears, notably over six hundred who managed to obtain North Carolinian citizenship and thus were exempt from deportation, and over four hundred Cherokee who hid in the remote Snowbird Mountains. Both groups later formed the beginning of the Eastern Band of Cherokees.

This map by James Mooney, dated 1900, shows the shrinkage of the Old Cherokee Country.

- Original Cherokee lands (blue line): Before European colonization they consisted of the entire state of Kentucky, save for the so-called Jackson Purchase (acquired in 1818 from the Chickasaw) in the far west of the present state; the bigger, eastern portion of Tennessee; the northern quarter of Alabama; the northern third of Georgia; the northwestern half of South Carolina; the western tips of North Carolina and Virginia; and a southwestern chunk of West Virginia.

- Cherokee boundary at close of Revolution (red line): Almost all of Kentucky is lost; the northern Cherokee border passes through the Cumberland Gap at the present-day state boundary tripoint between Virginia, Kentucky and Tennessee; so all Cherokee lands in West Virginia and Virginia are lost; a small tip of North Carolina remains; the Cherokee are almost completely withdrawn from South Carolina; in Georgia, the loss is minimal; in Alabama the southern border remains the same.

- Cherokee boundary at final cession (green line): The southern border remains the same, but in all other directions, the Cherokee Country has shrunk, now only occupying small parts of Alabama, Georgia and Tennessee.

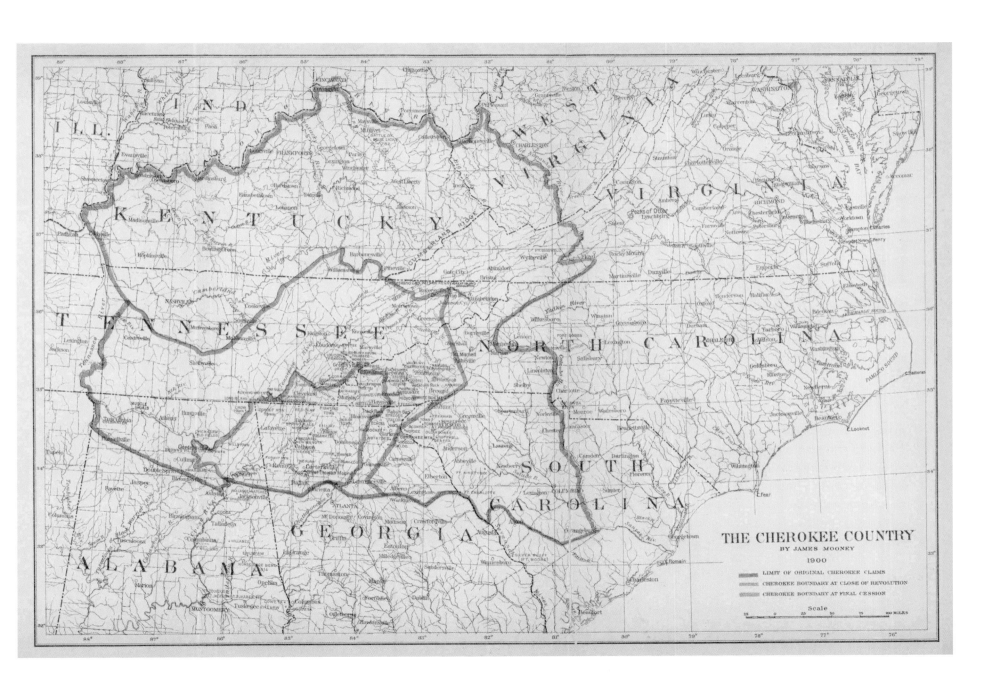

THE CHEROKEE COUNTRY

BY JAMES MOONEY

1900

	LIMIT OF ORIGINAL CHEROKEE CLAIMS
	CHEROKEE BOUNDARY AT CLOSE OF REVOLUTION
	CHEROKEE BOUNDARY AT FINAL CESSION

Scale

25   0   25   50   75   100 MILES

3

The Aroostook War
(Bloodless if You Don't Count the Pig)

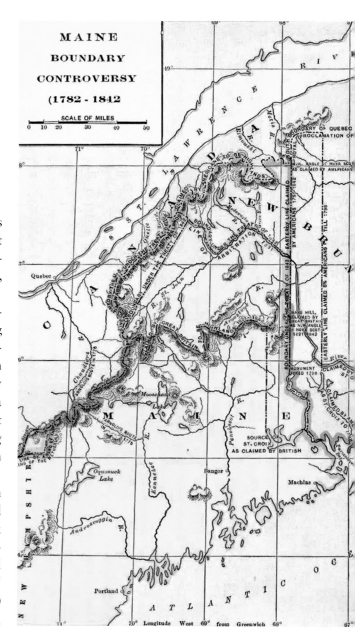

The wording of the 1783 Treaty of Paris that made American independence official proved a bit too vague where the U.S. northeastern border with the remainder of British North America was concerned. When the lumber-rich area became coveted by loggers from both sides of the ill-defined border, tensions rose.

Surveyors sent out in 1820 by Maine to mark out the new state's territory found Acadians on both banks of the St. John River. These French-speakers came from farther up north, and were British subjects. Land grants by Maine to American settlers led to disputes in which the king of the Netherlands was asked to arbitrate. In 1832, the U.S. Senate rejected the arbitration (which would have given the United States more territory than the eventual settlement in 1842).

In 1837, New Brunswick officers arrested a Maine government agent conducting a census in the disputed area. Maine dispatched two hundred "red shirts" north to confront the New Brunswick "blue noses"; Congress raised a 10,000-strong militia to support Maine's cause. The Americans built blockhouses to defend against a British invasion, but no actual fighting ever took place.

Legend has it the war had only one casualty—a Canadian pig killed after wandering over to the wrong side of the disputed border. Some sources relate that Private Hiram T. Smith, buried at Haynesville, was the only (human) casualty of the Aroostook War—a shaky claim, as no one seems to know what he died of. This frontier version of a *Sitzkrieg* lasted from 1838 to 1839 and became known as the Aroostook War.

In 1839, it was agreed that Congressman Daniel Webster (for the United States) and Lord Ashburton (for Britain) would work out a compromise. In 1842, the Webster-Ashburton Treaty stipulated that the United States should get over 7,000 square miles of the disputed area, and Britain almost 5,000 square miles. This map, dating from 1906, summarizes the main boundary disputes in the area between 1782 and 1842. Areas colored yellow and pink denote American and British (later Canadian) territory respectively, after the final settlement.

The legend on the red line reads, "Claimed by Great Britain to be highlands 1821–1842." The British based their claim for a border this far south on Mars Hill, shown directly west of Maine's straight north-south border with New Brunswick: "Mars Hill, claimed by Great Britain as Northwest Angle of Nova Scotia 1821–1842."

The legend on the northernmost part of the green line reads, "Claimed by Americans to be highlands." At the point where the line plunges south, it says, "Northwest Angle of Nova Scotia as claimed by Americans."

Between these two extremes lay the disputed area. In yellow is the "Line by award of the King of the Netherlands; line of arbitration 1831." The Webster-Ashburton Treaty mostly followed this line. The area to the north was lost to the United States, granting Britain the Halifax Road—an overland route between Lower Canada and Nova Scotia that was usable year round. A straight, dotted line to the south of the 1831 arbitration line denotes the "Boundary under the treaty of 1842." The territory in between could have been American, had the U.S. Congress adopted the Dutch arbitration proposal.

The small area caught between two borders in the southwest of this map later became the Republic of Indian Stream (see page 93). Dotted lines to the east of the actual border show earlier American claims.

4

World's Smelliest Border: The Limburg Split

For Americans, the word "Limburg" conjures up the mighty stink emitted by so-called Limburger cheese. In one of Mark Twain's stories, its smell is even mistaken for that of a corpse. As Superman dreads kryptonite, so the cartoon character Mighty Mouse can only be weakened by Limburger cheese.

In Europe, where the same cheese is known under several names, "Limburg" usually refers to one of four places. There is the German city of Limburg an der Lahn and the Belgian city of Limbourg (formerly the capital of an eponymous duchy). Furthermore, both Belgium and the Netherlands have a province called Limburg.

Those two provinces formed a patchwork of allegiances to dukes, counts and prince-bishops until they were unified into one administrative region by the conquering French in 1795. The French called this region the *département de la Meuse inférieure* (Department of the Lower Meuse) after the river that flows through it.

After the French defeat at Waterloo, this department became a province of the United Kingdom of the Netherlands (1815–30) and was renamed Limburg, after the former duchy—even though the territories of the old duchy and the new province barely overlapped, and the town of Limbourg wasn't included in the province of Limburg.

When Belgium split off from the Netherlands in 1830, Limburg—in the south and massively Catholic, like all the other rebellious provinces—was destined to go to Belgium in its entirety. But the Dutch mounted a successful military campaign, only called off under French pressure. While the Belgians occupied almost all of Limburg, Maastricht remained occupied by the Dutch general Dibbets. The ensuing stalemate is at the basis of the split.

Another reason was Prussia's unwillingness to tolerate a Belgian presence on the right bank of the Meuse, as Belgium was seen as a potential French ally and the prospect of a French army marching up so close to the German industrial heartland was viewed with alarm. In 1839, the Treaty of London confirmed the split, thus providing the Netherlands with its present-day borders and its southern "tail"—originally an anti-French buffer.

This stamp was issued by both Belgium and the Netherlands in 1989 at the 150th anniversary of the Treaty of London. That treaty lost Belgium half of Limburg and half of Luxembourg (which is similarly split in a Belgian province and an independent grand duchy, formerly in a personal union with the Dutch crown). It also required the new country to remain perpetually neutral (which it did in the run-up to both world wars—after the second one, Belgium was one of the founders of NATO).

The design on the stamp shows the territory of the Netherlands from the southern IJsselmeer to northern Belgium. The territory of Limburg west of the Maas River is colored a light orange and belongs to Belgium, while the territory north of the Netherlands-Belgium border and west of the Maas is a darker orange and belongs to the Netherlands.

Despite a separation that has lasted for about 175 years, a feeling of kinship between the Dutch and Belgian halves of Limburg persists, mainly due to similarities in culture and dialect (of Dutch) and by feelings of neglect by their respective "faraway" capitals, The Hague and Brussels.

5

An Accident of Nature, Geography and Man: Market Reef's Jigsaw Border

Measuring no more than 1,150 by 490 feet (350 by 150 m), Market Reef is the world's smallest sea island to be bisected by an international border. The tiny, uninhabited Baltic rock has a Swedish west and a Finnish east. But there's more that makes Märket (Swedish for "the Marker") *märklig* (strange).

On a visit to this flat, featureless fleck of land, you'd probably only notice the automated lighthouse and, if you didn't scare them off, the seals and birds that are the island's only inhabitants. But a map of Market Reef shows how the border, straight on either side of the island, morphs into a complex jigsaw inland, as if it's actually holding the two halves of the island together. Taking eight sharp turns where a single straight line would seem to suffice, the jagging borderline transfers equal parts of each half of the island to the sovereignty of the other half. This border is the result of triple chance: (a) a whim of nature, (b) an accident of geography and (c) human error.

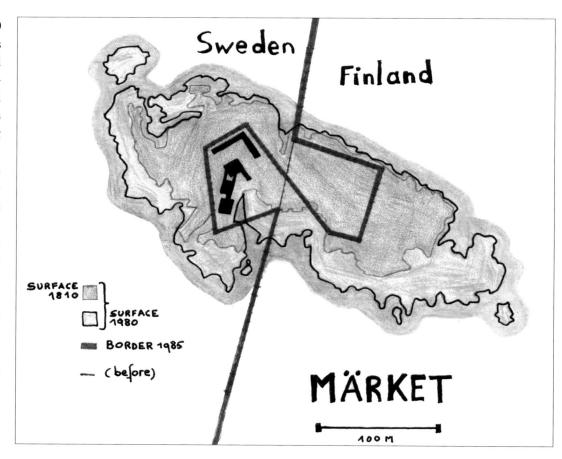

SURFACE 1810

SURFACE 1980

BORDER 1985

(before)

MÄRKET

100 M

(a) Not so very long ago, Market Reef didn't even exist. No mention is made of the island until the end of the eighteenth century, after which several shipwrecks are reported on its rocky shores. Pushed upward by local tectonics, the island has a rate of emergence of 5 millimeters per year, and grew 30 percent between measurements in 1810 and 1980.

(b) The borderline itself is a consequence of the Treaty of Fredrikshamn (1809), at which Sweden ceded Finland to Russia. The border was fixed equidistantly between the Swedish mainland and the Aland archipelago. When in 1811 it was discovered that the line ran straight through Market Reef, its eastern half became the Russian Empire's westernmost extremity.

(c) Market Reef adjoins the Understen-Märket Passage linking the Gulf of Bothnia to the Baltic Sea proper. To limit its danger to shipping, the Russians in 1885 constructed a lighthouse, which was manually operated until 1976. Only in the 1980s did a remeasurement of the border demonstrate that the lighthouse was actually located on the Swedish side. Both governments (Market Reef had passed from Russia to Finland at the latter's independence in 1917) rectified the problem by bending the border.

Because changing the sea boundary would have affected sovereignty over fishing zones, it was decided to exchange equal parts of territory on the island itself. The border, modified on August 1, 1985, now runs 490 meters across the island, as compared to the 100 meters of straight line before. Markers on the ground indicate where it makes a turn. The border will be resurveyed by both sides every twenty-five years; the next time will be in 2031. The lighthouse now is Finland's westernmost point.

Market Reef is the only place where Sweden shares a land border with the Aland Islands (a Swedish-language autonomous area of Finland). The island is a veritable border magnet. The Swedish side is shared by the communes of Norrtälje and Östhammar, which belong to the provinces of Stockholms län and Uppsala län respectively.

Radio amateurs consider Market Reef a "separate country" because of its special location, and a prized one due to its inaccessibility. Its prefix is OJ0. On regular island visits, radio enthusiasts make thousands of calls the world over.

6

May the Sauce Be with You: Battle Lines in the Barbecue Wars

"Tell me what you eat, and I'll tell you who you are": The famous quote by legendary gastronome Brillat-Savarin also applies very locally, to South Carolina, and very specifically, to barbecue sauce. All things barbecue are taken *very* seriously in the South, and the choice of barbecue sauce is the kind of subject on which strong opinions clash—with a vehemence usually reserved for religion, politics or sports. Some call it the barbecue wars—joking only slightly. Consider this comment by a whole-hog vinegar-style barbecue purist: "The mustard-based [sauce] is only fit for chicken or roadkill. . . . It is truly foul stuff."

This appetite-inducing piece of culinary cartography shows the Palmetto State divided into four regions, each with a different basis for their barbecue sauce:

1. Vinegar and pepper. Covers the eastern quarter of the state, and is a southern extension of a similar preferred barbecue sauce area in eastern North Carolina.

2. Tomato. Similarly connects to a North Carolinian sauce area; this Piedmont- or Lexington-style sauce is based on the eastern sauce—thin and vinegary, but with some tomato added.

3. Ketchup. Influenced by the preferred sauce in much of Georgia and most of the trans-Appalachian South, thick, sweet and ketchupy.

4. Mustard. Some sources say the mustard sauce of central South Carolina is unique to the state, others contend it extends throughout Georgia into Florida.

John Shelton Reed, a writer on southern culture and of *Holy Smoke: The Big Book of North Carolina Barbecue*, holds to the former opinion, and explains this presumed particularity by "the German names of the principal purveyors of mustard-based sauces. . . . It does seem that most are descended from the great 18th century wave of German immigrants to the Southern uplands."

A final word on that most appetite-inducing

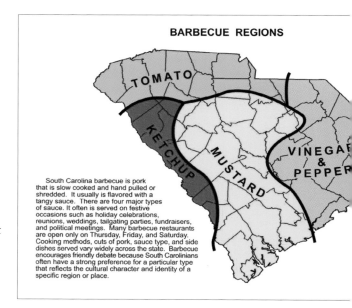

BARBECUE REGIONS

South Carolina barbecue is pork that is slow cooked and hand pulled or shredded. It usually is flavored with a tangy sauce. There are four major types of sauce. It often is served on festive occasions such as holiday celebrations, reunions, weddings, tailgating parties, fundraisers, and political meetings. Many barbecue restaurants are open only on Thursday, Friday, and Saturday. Cooking methods, cuts of pork, sauce type, and side dishes served vary widely across the state. Barbecue encourages friendly debate because South Carolinians often have a strong preference for a particular type that reflects the cultural character and identity of a specific region or place.

word *barbecue*—its etymology is not, as I always thought, French (from the roasting of wild boar snout to tail, or in French *barbe à queue*) but apparently it's an americanism, in fact a real southern word, derived from the New World Spanish expression *barbacoa*, itself taken from the Arawak language, where *barbakoa* means something like "wooden support beam"—the Arawaks' favorite instrument for their barbecues. History doesn't record which condiment *they* preferred.

X. EXCLAVES AND ENCLAVES

What islands are to geography, exclaves and enclaves are to cartography. Or something like it.

1

You Say Enclave, I Say Exclave: Let's Fence the Whole Thing Off

Enclaves are the islands of geopolitics—bits of national territory adrift from the motherland, surrounded by another country. But wait a minute, wasn't that what an *exclave* was? Well, enclaves are also exclaves . . . but not always.

These diagrams will explain most of the situations in which sovereign territories don't stay out of each other's way. But first let's get our definitions straight.

- An enclave (from the Latin *inclavatus*, "locked in") is sovereign territory—a part or the whole of a country—surrounded by another country;
- An exclave (from *exclavatus*, "locked out") is sovereign territory separated from the "mainland" by another country's territory.

The main difference is one of point of view: The enclave is defined by the territory that surrounds it, the exclave by the territory from which it is separated. Still confused?

Well, that's what these diagrams are for.

A. LLIVIA

This is your "classic" enclave, of the least confusing type, because it is also an exclave. The little village of Llivia, just north of the Pyrenees, is a Spanish exclave in France, and consequently an enclave of France. The 1695 Treaty of the Pyrenees ceded the villages around Llivia to France, but since Llivia was considered a city, it remained Spanish.

Other examples of type A enclaves are Büsingen, a German town surrounded by Swiss territory, and Ormidhia and Xylothimou, Greek Cypriot territory enclaved within the British Sovereign Base Area of Dhekelia.

B. JUNGHOLZ

The Austrian village of Jungholz is a pene-enclave in Germany (*paene* is Latin for "almost," also used in "peninsula") and consequently a pene-exclave of Austria. Jungholz is the purest of pene-enclaves, as it is con-

nected to its *Heimat* by a single point—the top of Mount Sorgschrofen.

C. ADEMUZ

The Spanish *comarca* (shire) of Ademuz is an exclave of the *comunidad* of Valencia, but it is not an enclave, as it is surrounded by more than one other entity: the Communities of Aragon and Castilla–La Mancha.

Another type C territory is Nakhchevan, an exclave of Azerbaijan, hemmed in by Armenia, Iran and Turkey—but not an enclave in any of those countries.

D. OCCUSI-AMBENO

Occusi-Ambeno (also spelled Oecussi-Ambeno) is part of the independent nation of East Timor, even though it is situated in the Indonesian west of that island. The difference between type A and this one is the sea: The mother country has access to her extraterritorial territory via the water, which also prevents it from being completely surrounded by other territory. Specialists therefore insist on calling these entities "fragments," but in general parlance they are referred to as enclaves and exclaves.

E. KALININGRAD

Similar to type C because it is surrounded by more than one country, and to type D because it also has access to the sea. So the Russian oblast (region) of Kaliningrad, on the Baltic Sea and surrounded by Poland and Lithuania, should be called a "fragment" of Russia. But because this sounds disrespectful enough to anger a resurgent Russia, and analogous to type D, Kaliningrad can be called an Russian exclave (though not an enclave).

F. LESOTHO

The Kingdom of Lesotho is completely surrounded by South Africa, and is therefore an enclave within South Africa. It is, however, not an exclave, as there is no other territory to be exclaved *from*. Other examples of sovereign states completely surrounded by another country are the Vatican and San Marino, both in Italy.

G. THE GAMBIA

The African nation of the Gambia is entirely encapsulated by Senegal, and would be a type F enclave if it did not have access to the sea. Purists don't consider type F enclaves to be "real" enclaves, but since they don't offer any cool alternative names, we might as well continue to do so.

    Another example of a type G "enclave" is Monaco, surrounded by France and the Mediterranean Sea.

H. SCHMALKALDEN/SUHL

Now this is really stretching it—and in

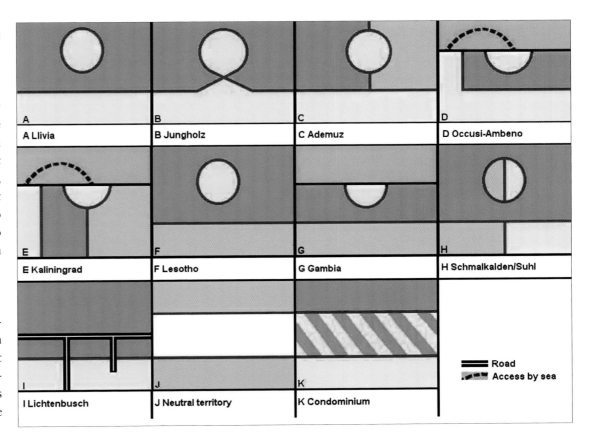

A Llivia
B Jungholz
C Ademuz
D Occusi-Ambeno
E Kaliningrad
F Lesotho
G Gambia
H Schmalkalden/Suhl
I Lichtenbusch
J Neutral territory
K Condominium

━━ Road
▄▀▄ Access by sea

fact, the only known example of type H is no longer in existence. In Germany after World War I, a Prussian exclave in the state of Thuringia consisted of parts of two different provinces, Saxony and Hesse-Nassau. Only on a state level would purists consider this an enclave (and an exclave), while on the provincial level both can still be considered exclaves, but not enclaves. If I prefer Llivia, it's not just because of the weather.

I. LICHTENBUSCH

Lichtenbusch (i.e., the green area hemmed in by the roads and the yellow bits) is only an enclave in a practical sense—it would be possible to get to the "mainland" without entering the other country, but that would involve crossing highways on foot, or clambering over fences.

J. NEUTRAL TERRITORY

A territory belonging to a government but that is de facto shared with other countries.

K. CONDOMINIUM

A joint or concurrent dominion of a territory by two or more states.

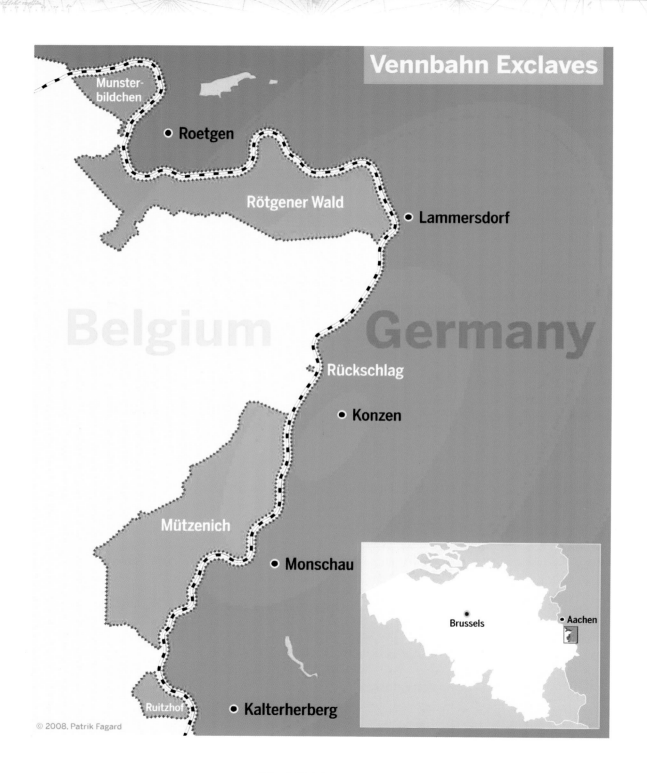

Vennbahn Exclaves

Munster-
bildchen

• Roetgen

Rötgener Wald

• Lammersdorf

Belgium   Germany

Rückschlag

• Konzen

Mützenich

• Monschau

Ruitzhof   • Kalterherberg

Brussels   • Aachen

© 2008, Patrik Fagard

2

The Vennbahn Complex: Five Little Germanies in Belgium

The Prussians opened the Vennbahn railway line in 1889 to transport coal and steel between Aachen and Luxembourg. At that time, the wild Hohes Venn (High Fens) area it ran through was entirely German. But after the Treaty of Versailles (1919) accorded about 350 square miles (900 sq. km) of German territory to the Belgians as part of Germany's reparations, a large tract of the Vennbahn became Belgian . . .

. . . except for the stretch between Roetgen and Kalterherberg, where the border and the railway line intertwined. To eliminate several border crossings that might obstruct the free flow of traffic, Belgium demanded sovereignty over the railway track itself.

On February 21, 1920, a special train rode the Vennbahn between Kalterherberg and Roetgen. The Inter-Allied Border Commission, consisting of a Brit, a Frenchman, an Italian and a Japanese, came to inspect the stretch of railway that Belgium demanded sovereignty over.

At the Monschau train station, they were met by thousands of protesters. But despite protests by the German government, Germany was not in a position to negotiate and Belgium got what it wanted: about 22 miles (36 km) of trackbed, winding its way through idyllic fen landscapes, in the process separating five small bits of Germany, to the west of the line, from the Fatherland.

Those enclaves are, north to south: Munsterbildchen, Rötgener Wald, Rückschlag, Mützenich and Ruitzhof. Rückschlag (literally "backlash") has the distinction of being Germany's smallest exclave: 1.5 hectares (3.7 acres).

The Mützenich enclave contains an entire village, possibly originally a Roman settlement. A Roman soldier in full armor was dug out of the fens in 1783—dead by then. His helmet now forms part of the communal coat of arms. Mützenich is a popular starting point for hikes into the High Fens.

Just inside the enclave is a large glacial erratic stone, called Kaiser Karls Bettstatt (Charlemagne's Bed). Legend has it that Charlemagne, having lost his way on a hunting party, had to spend the night here. When a servant presented him with a sleeping cap, Charlemagne replied, "Mütze nicht" ("No cap"), thus giving the village its name.

Belgian rail discontinued the Vennbahn line in 1989, after which a private, steam-powered train carried tourists on the picturesque line. When this also ended, in 2001, it was unclear whether the disused trackbed would remain Belgian territory or would have had to be handed back to Germany. In early 2008, the Belgian and German foreign ministries published a statement clarifying that the Vennbahn trackbed would remain under Belgian sovereignty, thus perpetuating the five German enclaves within Belgium. The track is now being converted into a bike trail.

3

Enclaves, Counterenclaves and a Dead Body: The Borders of Baarle

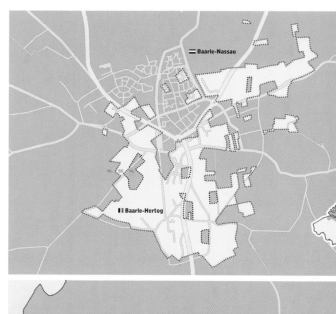

One of the world's unlikeliest enclave/exclave complexes is to be found on the Belgian-Dutch border, and is centered on Baarle. Essentially one town, and located entirely within the Netherlands, Baarle consists of two administrative units: the Dutch commune of Baarle-Nassau and the Belgian commune of Baarle-Hertog.

Baarle-Hertog (7.48 square kilometers; 2.88 square miles) consists of twenty-two separate enclaves in the Netherlands (H1–H22) and one piece of territory within Belgium proper (the little village of Zondereigen). Baarle-Nassau (76.36 square kilometers; 29.48 square miles) has eight enclaves (N1–N8), one situated within Belgium proper, seven within the Baarle-Hertog enclaves (these are counterenclaves). The situation goes back to feudal times, with Baarle-Hertog belonging to the Duke of Brabant (*hertog* is Dutch for "duke") and Baarle-Nassau owing allegiance to the Count of Nassau.

When Belgium gained its independence from the Netherlands in 1830, it proved impossible to reach a definitive agreement on the border in the area around Baarle. In the Maastricht Treaty of 1843, both governments opted to allocate nationality separately to each of the over 2,000 parcels of land in the 50 kilometers (31 miles) between border posts 214 and 215. These eventually coagulated into the enclave complex as it is today.

"Complex" is the right word: The border cuts through streets and houses throughout Baarle. This surreal situation is marked out on the ground, and has spawned a small tourist industry in the town. Some absurd effects of the bizarre Baarle borders:

- The "front door rule" determines that each house has to pay taxes in and get its utilities from the country in which its front door is located. This sometimes resulted in front doors being moved to enjoy a more favorable tax regime in the other country. To avoid confusion, all houses have their numbers in the national colors of either Belgium (black-yellow-red) or the Netherlands (red-white-blue). One house, where the border runs through the front door itself, has two addresses: Loveren 2 in Baarle-Hertog and Loveren 19 in Baarle-Nassau.

- The border cuts through houses, shops, pubs and the cultural center (which also has two addresses, one Belgian and one Dutch). In one pub, a billiard table straddles the border. It used to be that Dutch restaurants had to close earlier; in one eatery, this could be resolved by the clients moving to tables at the Belgian side.
- H22, at 2,632 square meters the smallest of all Baarle enclaves, was accorded to Belgium only in 1996, after it was discovered that it was the only one of the orginal parcels whose nationality had not been determined.
- A murder case in early 2008 was complicated by the fact that the body was discovered on the border itself, forcing detective teams from each country to look for clues in their half of the crime scene.

Color code for this map: Black indicates main border, red indicates Belgian enclaves inside the Netherlands, purple indicates Dutch enclaves within the Belgian ones, green indicates the only Dutch enclave within the Belgian "mainland."

4

Bubbleland, Not Far from Monkey's Eyebrow: The Kentucky Bend

Half man-made, half the Big Muddy's work, a 17.5-square-mile (45.5 sq. km) enclaved border irregularity bounded on three sides by an oxbow bend in the Mississippi and in the south by Tennessee is known as the Kentucky Bend, but its denomination is as unfixed as the river that created it. Alternate names are the New Madrid Bend, the Madrid Bend, the Bessie Bend and even "Bubbleland"—quite an image-provoking epithet; one involuntarily pictures Michael Jackson's monkey's own version of Neverland.

The U.S. Census doesn't count monkeys, however. According to the latest census poll in 2000, Bubbleland was home to seventeen Kentuckians, cut off from the mainland of their state by Missouri and Tennessee. Formally, their home is an exclave of Fulton County in Kentucky's extreme southwest. It is only reachable via Tennessee State Route 22.

The event that created Bubbleland was the New Madrid Earthquake, actually a series of earthquakes in late 1811 and early 1812 that each may have registered 8.0 on the Richter scale, making them the largest quakes in the contiguous United States. Not only flattening most of the town of New Madrid nearby in Missouri, the tremors—felt as far away as Connecticut—also shifted the course of the Mississippi.

This confounded the work of early surveyors plotting out the line that would mark the border between Kentucky and Tennessee. By 1812, they hadn't made it as far as the Mississippi. Later it turned out that their line cut right through the loop in the Mississippi created by the quakes, crossing the river twice.

This led to legal wrangling between Kentucky and Tennessee, for Kentucky had secured the Mississippi as its western border and thus claimed the westernmost point on the line. Tennessee held that it nevertheless had rights on the land contained in the loop. In fact, Tennessee administered Bubbleland as part of its Obion County until at least 1848, but eventually dropped its claim.

Much to its regret, one can imagine, as the fertile soil inside the loop proved extremely fertile cotton-growing land. The 1870 Census tallied more than three hundred residents, mostly cotton farmers. Interestingly, Bubbleland has two other claims to fame:

- From February 28 to April 28, 1862, the area was the location of the Battle of Island Number Ten between Union and Confederate forces in the American Civil War. The battle, which involved ironclad ships, was won by the Union side and opened up the Mississippi farther south, eventually leading to the capture of Memphis by Northern troops. Island Number Ten has since eroded away (although Island Number Nine still remains).
- In *Life on the Mississippi* (1883), Mark Twain describes a vendetta lasting sixty-odd years between the Darnell and Watson families living in Bubbleland: "Both families belonged to the same

church. . . . They lived each side of the line, and the church was at a landing called Compromise. Half the church and half the aisle was in Kentucky, the other half in Tennessee. Sundays you'd see the families drive up, all in their Sunday clothes, men, women, and children, and file up the aisle, and set down, quiet and orderly, one lot on the Tennessee side of the church and the other on the Kentucky side; and the men and boys would lean their guns up against the wall, handy, and then all hands would join in with the prayer and praise; though they say the man next the aisle didn't kneel down, along with the rest of the family; kind of stood guard."

Whether this blood feud is in some way responsible for the thinning of the population of Bubbleland could not be ascertained.

5

Madha and Nahwa: The Omelet-Shaped Enclave Complex

These names sound like they're out of the Arabian Nights, and that's not far off. Madha and Nahwa are the names of two obscure but cartographically intriguing territories on the Arabian Peninsula. Together they form a type of enclave/exclave complex that I would like to call the Omelet.

The territories are situated on the horn of the peninsula, the one pointing at Iran and prevented from touching it by the Strait of Hormuz. The tip of the horn is an enclave of Oman, the mainland of which is farther to the south. On this map, Musandam (i.e., that little piece of Oman on the strait) is not separately named, although several towns are (including Kumzar, on its own oil spill–shaped peninsula) and one interesting natural feature is pointed out: the Jabal al Harim, surprisingly high at 2,087 meters (6,847 feet). One always thinks of Arabia as flat, sandy desert—apparently not entirely correct.

Below this unnamed enclave, the territory of the United Arab Emirates fans out south and west, separating it from Oman proper—and giving the UAE sea access to the Indian Ocean in the process. On the other UAE shore, the one on the Persian Gulf, lies the glittering metropolis of Dubai, currently courting world attention with huge building projects.

On the same latitude as Dubai, but closer again to the Indian Ocean, lie Madha and Nahwa. Madha (75 square kilometers; 29 square miles) is an Omani enclave within UAE territory, while Nahwa in turn is UAE land, completely surrounded by Oman. This rather complex border situation was definitively demarcated in 1969. Nahwa consists of only forty-odd houses, while most of Madha is uninhabited, making the necessity for this particular delineatory arrangement even more mysterious.

6

Cooch Behar: The Mother of All Enclave Complexes

Nowhere is an international border—or any border, for that matter—more complex than between Cooch Behar, a district of the Indian state of West Bengal, and Bangladesh. You'd need a map the size of a football field to do justice to the intricacy of this, the world's largest enclave complex. But a map of that size would make it difficult to get an overview—not to mention that it wouldn't fit in the glove compartment. A good compromise is this map by Dr. Brendan Whyte, who wrote a thesis on the enclaves, counterenclaves and counter-counterenclave (the world's only one!) of Cooch Behar.

Cooch Behar, formerly an independent Indian principality, possesses 106 exclaves in Bangladesh, totaling 69.6 square kilometers (26.9 square miles). Of those, three are counterenclaves and one the aforementioned counter-counterenclave. The biggest Indian enclave is Balapara Khagrabari (25.95 square kilometers; 10 square miles), the smallest Panisala (1,093 square meters; 11,765 square feet).

Conversely, Bangladesh possesses ninety-two exclaves inside India, comprising 49.7 square kilometers (19.1 square miles). Of these, twenty-one are counterenclaves. The largest Bangladeshi exclave is Dahagra-Angarpota (18.7 square kilometers; 17.2 square miles), the smallest is the counterenclave Upan Chowki Bhajni (2,870 square meters; 30,892 square feet). At the start of the twenty-first century, estimates for the population of all Indian and Bangladeshi enclaves together ranged up to 70,000.

Dr. Whyte's map shows three concentrations of enclaves:

First, and westernmost: a mainly Indian archipelago of enclaves, inside Bangladesh. The enlarged area at lower left shows the two largest Indian enclaves, with a few Bangladeshi dots (counterenclaves) inside them. The southern area of the eastern enclave of Balapara Khagrabari is again enlarged to show one of these Bangladeshi counterenclaves, Upanchowki Bhajni 110. Surrounded by this area is Dahala Khagrabari 1/51—the world's only counter-counterenclave.

Secondly, and centrally: a mixed Indo-Bangladeshi archipelago, consisting of a number of Indian enclaves inside the administrative area of Patgram—which itself is a Bangladeshi protrusion into Cooch Behar (but contiguous with Bangladesh proper). To the east, north and west of Patgram, and therefore within India, are a number of Bangladeshi enclaves, the largest of which, Dahagram-Angarpota, has been agreed will remain Bangladeshi even when (or rather, if) the other enclaves are all eventually exchanged. The Tin Bigha corridor, measuring a mere 178 by 85 meters (584 by 279 feet), leased in perpetuity to Bangladesh but open daylight hours only, connects it to Bangladesh proper.

Lastly, to the east, and again shown enlarged: a mixed archipelago, but more spread out than the previous two, with Indian exclaves inside the Bangladeshi districts of Lalmonir Hat, Fulbari, Kurigram and Bhurunghamari and Bangladeshi exclaves inside the Indian subdistricts of Dinhata, Cooch Behar proper and Tufanganj.

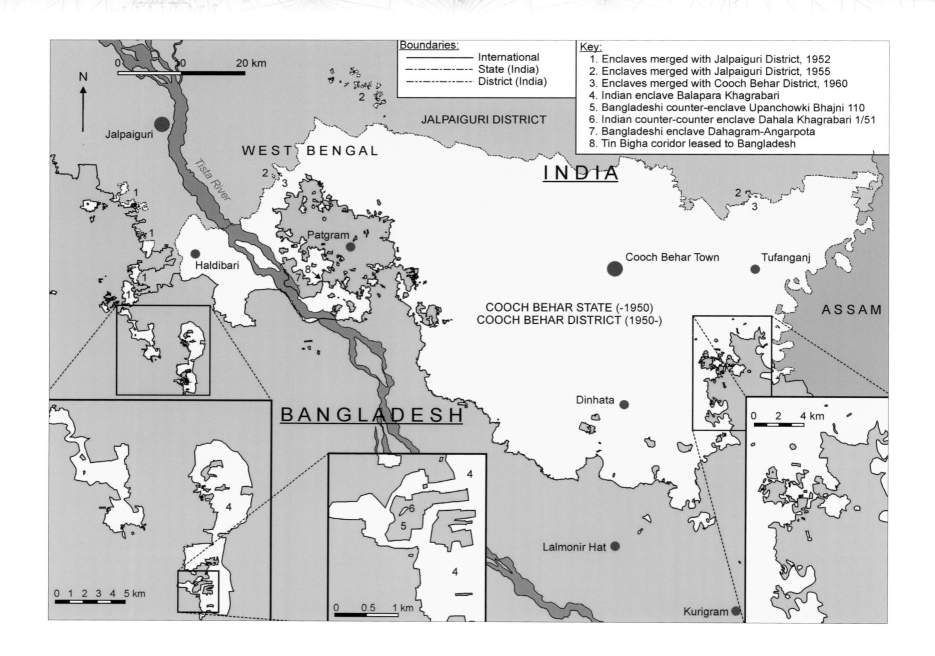

Boundaries:
———————— International
—·—·—·— State (India)
—··—··—·· District (India)

Key:
1. Enclaves merged with Jalpaiguri District, 1952
2. Enclaves merged with Jalpaiguri District, 1955
3. Enclaves merged with Cooch Behar District, 1960
4. Indian enclave Balapara Khagrabari
5. Bangladeshi counter-enclave Upanchowki Bhajni 110
6. Indian counter-counter enclave Dahala Khagrabari 1/51
7. Bangladeshi enclave Dahagram-Angarpota
8. Tin Bigha coridor leased to Bangladesh

JALPAIGURI DISTRICT

WEST BENGAL

INDIA

Jalpaiguri

Tista River

Patgram

Haldibari

Cooch Behar Town

Tufanganj

ASSAM

COOCH BEHAR STATE (-1950)
COOCH BEHAR DISTRICT (1950-)

BANGLADESH

Dinhata

0   2   4 km

Lalmonir Hat

0 1 2 3 4 5 km

0   0.5   1 km

Kurigram

The enclave complex originated after a treaty in 1713 between the Mughal Empire and the Cooch Behar Kingdom reduced the latter's territory by one-third. The Mughals didn't manage to dislodge all Cooch Behar chieftains from the territory they had gained; at the same time, some Mughal soldiers retained lands within Cooch Behar proper while remaining loyal to the Mughal Empire. This territorial "splintering" was not so remarkable in the context of that time: The subcontinent was extremely fragmented, most enclaves were economically self-sufficient and the fragmentation caused no significant border issues, as Cooch Behar was nominally tributary to the Mughals anyway.

In 1765, the British East India Company, seizing control of the Mughal territory, was surprised to discover that "by some unaccountable accident," there were extraterritorial dots of Cooch Behar within its territory, and vice versa. Those enclaves were used as sanctuary by "public offenders" fleeing the police. In 1947, the formerly Mughal territories of British India were split between India and Pakistan, and Cooch Behar found itself straddling the divide between India and the eastern half of Pakistan. Cooch Behar acceded to India only in 1949, as one of the last of the 600-odd preindependence princely states to do so, and became a simple district of West Bengal the following year. In 1971,

East Pakistan gained independence as Bangladesh.

Remarkably, the enclave complex survived all these changes of sovereignty on both sides of the border, although its intra-Indian intricacies were somewhat simplified: When Cooch Behar, Assam and West Bengal became parts of India, fifty-something enclaves between them were rationalized away.

Attempts in 1958 and 1974 to do the same across the international border and exchange enclaves between Bangladesh and India proved more elusive, even though these were more problematic and unworkable than the all-Indian ones. The border situation has often made it impossible for people living in the enclaves to legally go to school, to hospital or to market. Complicated agreements for policing and supplying the enclaves had to be drawn up (a 1950 list of products that could be imported into the enclaves contained such items as matches, cloth and mustard oil).

In a classic catch-22 situation, residents of enclaves need visas to cross the other country's territory toward the "mainland," but since there aren't any consulates in the enclaves, they should go to one in the "mainland"— which they can't because they don't have a visa. Illegal border crossings are frequent, but dangerous—a number of transgressors have been shot by border guards. Furthermore, the enclaves remain a haven for criminals who are

then immune from the justice system of the country surrounding the enclave, exactly as was lamented by the British on their takeover. These and other problems have rendered the enclaves pockets of lawlessness and poverty even compared to their already relatively poor motherlands.

Since the issues of sovereignty, territorial integrity and especially the unwillingness to let the other side seem to "win" is so sensitive for both India and Bangladesh, the Cooch Behar enclave complex probably isn't going to disappear anytime soon. There is one example of progress, however: the Tin Bigha corridor, connecting the Bangladeshi enclave of Dahagram-Angarpota with its "mainland," although it took twenty years to happen, met heavy opposition and cost people's lives.

Though the Cooch Behar enclave complex may be the world's most intricate enclave complex, it remains quite obscure abroad— cartographic anomalies in the Western world tend to attract more attention. One oblique reference in (Western) popular culture might be in Bertolt Brecht's "Cannon Song" (from *The Threepenny Opera*): "The troops live under the cannon's thunder / From the Cape to Cooch Behar." Brecht used the name of the (then still) princely state to symbolize the far end of the world; but he was, among many other things, a cartophile. He may have been drawn to Cooch Behar precisely by its most irregular border.

XI. A MATTER OF PERSPECTIVE

North isn't up, the time is not now, and things generally aren't what they seem—or seem to be something they're not.

1

Subcontinent with a Twist: Sri Lanka on Top!

Funny how something as arbitrary as map orientation can skew the perception of countries. On this map, the Indian Subcontinent is shown "upside down": South is top of the map, north is at bottom. Consequently, east is left and west is right (otherwise the map would be mirrored). In speleological terms, the Indian Subcontinent has changed from a stalactite to a stalagmite.

The island nation of Sri Lanka (formerly known as Ceylon) dominates the top of the map—and suddenly seems more like the jewel in the crown of the subcontinent, less like an appendage of its larger neighbor India.

Bangladesh seems more like the sinkhole for some big rivers than it actually is, and India appears to have a viselike grip on the Chinese territory of Tibet: The trans-Bangladeshi part of India doesn't so much seem to be a drifting piece of territory as one half of the aforementioned vise.

But that's just my utterly fanciful interpretation. The map was devised by *Himal*, a weekly magazine in Kathmandu, Nepal, that seeks to restore some of the historical unity of "Southasia" (*sic*): "We believe that the aloof geographical term 'South Asia' needs to be injected with some feeling. 'Southasia' does the trick for us."

On the "right side map" itself, *Himal* explains: "This map of South Asia may seem upside down to some, but that is because we are programmed to think of north as at the top of the page. This rotation is an attempt by the editors of *Himal* (the only South Asian magazine) to reconceptualize 'regionalism' in a way that the focus is on the people rather than the nation-states. This requires nothing less than turning our minds downside up."

Similar downside-up world maps exist, showing Australia on top of the world instead of dangling off toward the distant south. There is no good reason why north should be "up," except convention (which seems to have been started by Ptolemy, the second-century Egyptian astronomer and cartographer). In the Middle Ages, incidentally, many maps were made with the east on top (hence the word "orientation").

2

Your Antipodes Most Likely Have Fins

Imagine that you could drill a hole straight through the Earth. Suspend your disbelief for a moment, ignoring the molten core that would fry you. Where would you end up?

In geographical coordinates, the answer is quite simple: If the coordinates (longitude and latitude) of a point on the earth's surface are (x, y), then the coordinates of the antipodal point can be written as $(x \pm 180°, -y)$. So the latitudes are numerically equal, but one is north and the other south. And the longitudes differ from each other by 180 degrees. Plus or minus, it doesn't really matter in which direction you count those 180 degrees, as either way will lead you to the same point (a circle having a circumference of 360 degrees).

An example. If you start out at, say, 46.95 degrees longitude west and 39.00 degrees latitude north, after you've dug through the earth's core you'll end up at longitude 133.05 degrees east (133.05 being the result of 180.00 − 46.95) and latitude 39.00 degrees south.

Only, for most people, the place where you'll end up won't be land, but water. The

oceans cover about 70 percent of our planet's surface. Your antipodes (a Greek word translatable as "those whose feet are on the opposite side") mostly don't have feet, but fins. The overlap of land is surprisingly small.

This is reminiscent of the movie *The China Syndrome*, the title of which refers to the idea that if you dig a hole through the Earth starting in the United States, you end up in China. This map shows it ain't so.

- In fact, only a little bit of China overlaps— and with the southern part of South America. Funnily enough, the good people of Argentina seem to have taken this into account when naming the province of Formosa, which is the antipode of Taiwan, the island off the Chinese coast formerly known as . . . Formosa.

- Indonesia overlaps with parts of Ecuador, Colombia, Venezuela, Guyana, Suriname (also a former Dutch colony), French Guiana and Brazil. Malaysia and Vietnam are also partly Peruvian antipodes. The Philippine main island of Luzon is partly in Bolivia, the rest is in Brazil. The Falkland Islands lie on the Russian-Chinese border.
- There's almost no overlap in North America, except for the northern parts of Canada, Greenland and Alaska with Antarctica and a tiny speck straddling the Montana (U.S.)-Saskatchewan (Canada) border with Kerguelen Island.
- Africa is antipodes-free, except for some Pacific islands (Hawaii overlaps with northern Botswana, for example).
- In Europe, only the Iberian Peninsula

overlaps with a significant chunk of land (New Zealand).

Some other interesting antipodes (or near-antipodes):

- Seville (Spain) and Auckland (New Zealand)
- Hamilton (Bermuda) and Perth (Australia)
- Taipei (Taiwan) and Asunción (Paraguay)
- Jakarta (Indonesia) and Bogotá (Colombia)
- Punta Arenas (Chile) and Irkutsk (Russia)
- Mecca (Saudi Arabia) and Avarua (Cook Islands)
- Cherbourg (France) and . . . the Antipodes Islands (New Zealand)

3

Go with the Flow: The Norwegian Drop

This remarkable painting was made by the Norwegian artist Rolf Groven as a poster proposal for Norway's pavilion at the World Exhibition in Seville, Spain, in 1992. The title is *Den Norske Dråpen* (The Norwegian Drop). Water is very significant indeed for Norwegians, as hydroelectric power produces 98.5 percent of the electric power generated in-country—this in spite of Norway's huge North Sea oil reserves, which consequently are exploited mainly for export.

"This painting is aimed at visualizing how this energy source is entirely renewable and is a result of Norway's distinct geography," Mr. Groven states on his Web site. And it does just that:

- Norway is a foaming mass of water gushing down a rocky mountainside that to the right looks like the rest of Scandinavia.
- A nice touch: Iceland is formed by a . . . spot of ice on the side of the mountain towering over the landscape.
- Rivulets of water form the boundaries

of Finland and Sweden, Russia's Kola peninsula is defined by the stagnant pond next to it.

- The "head" of Norway at its southern end is a waterfall, perpetually showering Denmark's Jutland peninsula with crystal-clear Norwegian water.
- That water flows on to etch the edges of Europe out of its rocky landscape—clearly a reference to what the northern desolation of Norway must look like.
- A road winding down through northern Germany, past the Benelux countries and via France leads to Italy, transformed into a road, splitting near where Italy's real-life heel and toe are situated. A road sign invites us to take the other direction, up toward Norway.
- To the left, a salmon and the British Rocks are floating quite mysteriously above the water—perhaps all three of them have just leapt up out of the mountain stream.
- It has been suggested that the mysterious lights peering out of the rock in the

western Mediterranean might be the eyes of Nix, a shapeshifting water spirit of Germanic lore.

- On closer—or rather, farther—inspection, the landscape is situated not in a crystal ball, but in a lightbulb—appropriately referring to Norway's sensible exploitation of its renewable hydroelectric resources.

Rolf Groven (born 1943) studied art in Norway and architecture in Iran and worked as a builder, sailor, architect and teacher before settling on painting and illustrating as his main occupation.

4

Humboldt's Imaginary Mountain: The Beginning of Geobotany

Wilhelm von Humboldt (1767–1835) was a philosopher, linguist and diplomat, friend of Goethe and Schiller, founder of Humboldt University in Berlin, designer of the Prussian education system (later copied by Japan and the United States), groundbreaking researcher into the nature and origins of Basque and other languages, formulator of a theory of linguistic relativity a century before the Sapir-Whorf hypothesis did the same . . . And Wilhelm is the lesser famous of two brothers.

Said Schiller of the Humboldts, "Alexander impresses many, particularly when compared to his brother—because he shows off more!" Whereas Wilhelm sought knowledge and found fame inside the ivory towers of academia, Alexander (1769–1859) went out into the world to explore and taxonomize the natural world. The mere list of stuff named after this Humboldt hints of a life more adventurous than most action movies. In the United States alone, thirteen towns, three counties and two other locations are named after him. Mountain ranges from China to the Antarctic carry his name, as well as ships, an asteroid, plants, parks, a location on the moon and animals—species of bat, skunk, monkey, dolphin, penguin and (he won't be too pleased with this) a sucker (actually a freshwater fish)—never mind the Humboldt yeast.

Humboldt was the first to scientifically and exhaustively describe the flora and fauna of Latin America (1788–1804), where he discovered electric eels, climbed to the record altitude of 19,286 feet (5,878 m), discovered a connection between the Orinoco and Amazon river systems, described the fertilizing properties of guano, was the first to remark on the periodicity of the Leonid meteor shower and much, much more. "Von Humboldt has done more for America than all its conquerors," said Simón Bolívar. "He is the true discoverer of America."

Humboldt was also the founder of a discipline called botanical geography, or geobotany. In South America, he observed how physical conditions such as altitude influenced the geographic distribution of plants. This map, by Joseph Meyer (1860), describes (in German) "Humboldt's plant regions in southern America, after their elevation above sea level."

The map, taken from something called *Meyer's Handatlas* (why the hand? Was this the nineteenth-century equivalent of a palmtop?) shows a generic mountain, its altitudes indicated in Rhinelandic feet (1 equals 1.02 "regular" feet; left) and meters (right).

On either side of the mountain, some comparative heights are offered—five American altitudes, and for the benefit of the mainly European audience, three well-known Old World heights. In ascending order:

Vesuvius (Italy): 1,281 m (4,203 feet)
Quito (Ecuador): 2,850 m (9,350 feet)
Teyde (Spain): 3,718 m (12,198 feet)
Mont Blanc (France): 4,810 m (15,781 feet)
Popocatépetl (Mexico): 5,426 m (17,802 feet)
Orizaba or Citlaltépetl (Mexico): 5,636 meters (18,490 feet)
Cotopaxi (Ecuador): 5,897 m (19,347 feet)
Chimborazo (Ecuador): 6,268 m (20,565 feet)

Also mentioned, by the way, is the altitude achieved in 1735 by Bouguer and De La Condamine of the French geodetic expedition to South America as they reached the summit of the Corazón (4,970; 15,715 feet). The expedition was crucial in determining the exact length of a degree at the equator—and might have influenced the naming of the then Spanish province of Quito when it gained independence as Ecuador (1820).

Dotted lines indicate the lower and upper boundaries (4,000–9,000 feet) of the *Cinchonae*, a genus of about twenty-five evergreen shrubs and trees (5–15 m), all native to South America. *Cinchonae* are noted for their medicinal properties; the bark is the source of quinine and other alkaloids. Legend has it that taxonomist Linnaeus named the genus after the wife of a Spanish viceroy, the countess of Chinchon, who survived a near-fatal malarial attack thanks to this indigenous medicine, later also referred to as Jesuit's bark. In large doses, *Cinchonae* can be lethal. Not coincidentally, experimentation with these plants led to the birth of the homeopathic movement.

Straddling the lower border of the *Cinchonae* is the region of palm trees and Scitamineae, a group of flowering plants (up to 2,700 feet), above is the region of treelike ferns (1,200 to 4,500 feet) and the upper boundary of mimosa. On either side of the ravine to the right is the region of the *Coperniciae* (5,000 to 8,200 feet), a genus of twenty-four palms endemic to South America and the Caribbean, noted for their fanlike leaves, in some species covered in a thin layer of wax (hence the German name *Wachspalm*).

No specific altitude accompanies the treeline (*Obere Gränze der Bäume*), to the left and upward from the *Coperniciae*—but it's almost the same height as Quito. A bit higher still is the region of the *Barnadesia* and *Duranta*, a collection of flowering plants and shrubbery respectively. To the left is the region of *Wintera* and *Escallobia*, and higher the region of the *Chuquiraga*, *Gentian* and *Frailexon* (6,000 to 12,500 feet), with the plant world definitely thinning out to Gramineae (i.e., grasses) at 12,000 to 14,000 feet.

The top of the vegetal altitude chain is formed by the region of lichen and *Umbilicariae* (a genus of lichen known as rock tripe—used as food in extreme cases, and in oriental cuisine).

The whole side of the mountain, sliced off as if it was a giant cake, is covered with the exotic names of individual species, from "large pepper trees" (inexplicably in English on this mainly German/Latin map) to *Lobelia nana*, *Craniolaria* and *Jacquinia*—as if Humboldt's imaginary mountain doubles as a reservation for pretentious first names whose time has not yet come.

5

One Size Fits All: The Equinational Projection

Unfortunately, *Globehead! Journal of Extreme Cartography* was a rather short-lived grad school magazine at Penn State (only two issues in 1994), otherwise we might have seen some more strange maps like this one. This "equinational projection" goes where no Mercator or Peters projection ever went, and shows a world in which every country is the same size. A world that is a little different from ours:

- The American continent, especially its northern half, is covered by relatively few states, resulting in an atrophied New World—except for the Caribbean, where all those tiny island nations now each occupy the same space as giants like the United States and Canada.
- Europe, its territory littered with lots of states, medium and small (compared to America), holds a dominating position. Russia (number 138) is a mere appendage of Europe.
- Africa, long squeezed and thereafter

painfully stretched by the aforementioned Mercator and Peters projections, now holds what at first glance seems the largest block of nations.

- Asia consists of a few very large countries (Russia, China, India, Kazakhstan, Pakistan), which accounts for its relatively small size. This constrasts to almost any other projection, be it size, population or economic growth.
- Australia and New Zealand are the most visible constituents of Oceania, except on this map, where all the Pacific island nations figure more prominently than usual.

Note that the map is almost a decade and a half old, and thus doesn't include the world's newer nations (like East Timor, or Montenegro and Kosovo, the latest chips off the old ex-Yugoslav block).

And while an equinational projection might seem devoid of any practical use, it could, as has been pointed out, be used to

predict the strength of regional voting blocs when the UN General Assembly puts something to the vote.

"The equinational projection was invented by my friend Catherine Reeves for *Globehead!*" writes Jeremy W. Crampton, editor of *Cartographica* and associate professor of geography at Georgia State, who sent this map in. He kindly explains the cryptic acronym IASBS: International Association for the Study of Big Science. Ms. Reeves modestly adds that "this projection was commissioned by Mr. Scott Fares, Scientist #1 of the IASBS. I merely executed his concept."

Mr. Fares points out one other interesting aspect of this map: "Speed of update was a design objective as the USSR was closing shop during the weeks this was built. As one of the authors of the IASBS project, I am proud to recall that the frequent editions of this map were likely the first accurate political maps published for each of the breakaway republics, since it did not have to be concerned with disputed borders."

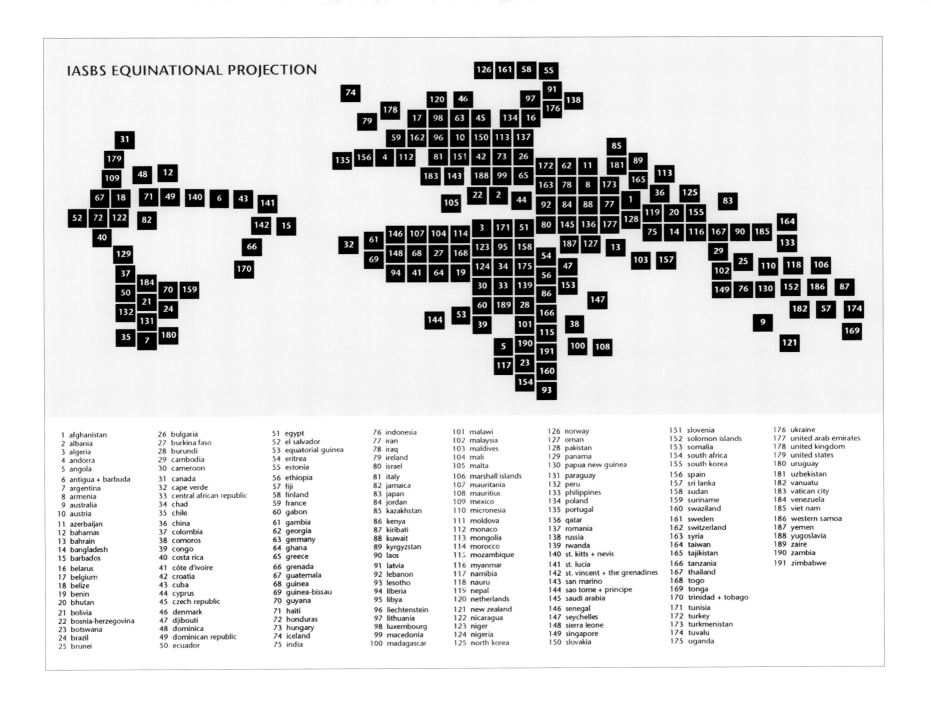

IASBS EQUINATIONAL PROJECTION

1 afghanistan
2 albania
3 algeria
4 andorra
5 angola
6 antigua + barbuda
7 argentina
8 armenia
9 australia
10 austria
11 azerbaijan
12 bahamas
13 bahrain
14 bangladesh
15 barbados
16 belarus
17 belgium
18 belize
19 benin
20 bhutan
21 bolivia
22 bosnia-herzegovina
23 botswana
24 brazil
25 brunei

26 bulgaria
27 burkina faso
28 burundi
29 cambodia
30 cameroon
31 canada
32 cape verde
33 central african republic
34 chad
35 chile
36 china
37 colombia
38 comoros
39 congo
40 costa rica
41 côte d'ivoire
42 croatia
43 cuba
44 cyprus
45 czech republic
46 denmark
47 djibouti
48 dominica
49 dominican republic
50 ecuador

51 egypt
52 el salvador
53 equatorial guinea
54 eritrea
55 estonia
56 ethiopia
57 fiji
58 finland
59 france
60 gabon
61 gambia
62 georgia
63 germany
64 ghana
65 greece
66 grenada
67 guatemala
68 guinea
69 guinea-bissau
70 guyana
71 haiti
72 honduras
73 hungary
74 iceland
75 india

76 indonesia
77 iran
78 iraq
79 ireland
80 israel
81 italy
82 jamaica
83 japan
84 jordan
85 kazakhstan
86 kenya
87 kiribati
88 kuwait
89 kyrgyzstan
90 laos
91 latvia
92 lebanon
93 lesotho
94 liberia
95 libya
96 liechtenstein
97 lithuania
98 luxembourg
99 macedonia
100 madagascar

101 malawi
102 malaysia
103 maldives
104 mali
105 malta
106 marshall islands
107 mauritania
108 mauritius
109 mexico
110 micronesia
111 moldova
112 monaco
113 mongolia
114 mozambique
115 mozambique
116 myanmar
117 namibia
118 nauru
119 nepal
120 netherlands
121 new zealand
122 nicaragua
123 niger
124 nigeria
125 north korea

126 norway
127 oman
128 pakistan
129 panama
130 papua new guinea
131 paraguay
132 peru
133 philippines
134 poland
135 portugal
136 qatar
137 romania
138 russia
139 rwanda
140 st. kitts + nevis
141 st. lucia
142 st. vincent + the grenadines
143 san marino
144 sao tome + principe
145 saudi arabia
146 senegal
147 seychelles
148 sierra leone
149 singapore
150 slovakia

151 slovenia
152 solomon islands
153 somalia
154 south africa
155 south korea
156 spain
157 sri lanka
158 sudan
159 suriname
160 swaziland
161 sweden
162 switzerland
163 syria
164 taiwan
165 tajikistan
166 tanzania
167 thailand
168 togo
169 tonga
170 trinidad + tobago
171 tunisia
172 turkey
173 turkmenistan
174 tuvalu
175 uganda

176 ukraine
177 united arab emirates
178 united kingdom
179 united states
180 uruguay
181 uzbekistan
182 vanuatu
183 vatican city
184 venezuela
185 viet nam
186 western samoa
187 yemen
188 yugoslavia
189 zaire
190 zambia
191 zimbabwe

6

Courses, Countries and Comparative Lengths: The World's Principal Rivers

At first glance, this map looks like an off-center representation of a polar region. But what would be the point of a compass rose, centered on the pole itself? One point too many at least, because it wouldn't be able to point north (or south, respectively).

Also, the distance of the lands surrounding the "pole" from that center is too uniform. Nature abhors a vacuum, but also anything too neatly symmetrical. No, this is an imaginary country, thought up by a nineteenth-century cartographer.

The circular expanse is a fictional sinkhole for the world's principal rivers, brought together as if around a campfire, better to compare not only their courses and lengths, but also their orientations—for each estuary is placed in a realistic relation to the compass rose. What a curious meeting this United Nations of rivers must be, what strange (river)bedfellows are met!

The Siberian river Lena, exactly to the south of the wind rose, is often frozen solid. Its neighbor, the sweltering Nile, is shown with its delta and its major port city Alexandria and with the capital city Cairo upstream. Not far away to the right are Antwerp, Cologne and Utrecht, cities on the lower reaches of the Scheldt, Meuse and Rhine. Then a succession of the Yenisei, Garonne, Seine and Vistula rivers places Paris next to Warsaw, and Toulouse close to Kraków.

Tiber and Don flow fraternally westward, placing Rome at the edge of the Sea of Azov. Other unlikely neighbors are Kherson on the Dnieper, New York on the Hudson and New Orleans on the Mississippi—not to mention Buenos Aires on the Rio de la Plata and Basra on the confluence of Tigris and Euphrates. The suburbs of Edinburgh on the Forth and London on the Thames probably overlap. Nottingham, on the tiny river Trent, is just downstream from Québec, bordering the majestic St. Lawrence River.

Since not all of the earth's principal rivers fit around the main "sinkhole," two others at the top left- and right-hand side of the page collect a few others. Concentric circles, each 200 miles apart, indicate the distance of each river from the sea. To mitigate somewhat the confusion caused by placing these disparate streams together, the Asian rivers are represented by solid lines, European ones by dotted lines, American rivers by-lines and African ones by alternating dots and lines.

The legend at the lower left-hand side indicates the length of the rivers, the longest ten being (in English miles):

1. Mississippi (from the source of the Missouri): 3,500
2. Amazon: 3,200
3. Yenisei (from the source of the Tula): 2,900
4. Obi: 2,800 (subsequently known as the Ob)
5. Nile: 2,750

6. Yangtze: 2,700
7. Lena: 2,500
8. Huang Ho: 2,400
9. Yoliba and Niger: both 2,300

This map was made in the year 1834, when the farthest sources of many of the rivers portrayed here had yet to be discovered. The top ten of longest rivers (in miles) has since been reshuffled:

1. Nile: 4,135
2. Amazon: 3,980
3. Yangtze: 3,917
4. Mississippi/Missouri: 3,902
5. Yenisei: 3,445
6. Huang Ho: 3,398
7. Ob: 3,364
8. Congo: 2,922
9. Amur: 2,763
10. Lena: 2,736

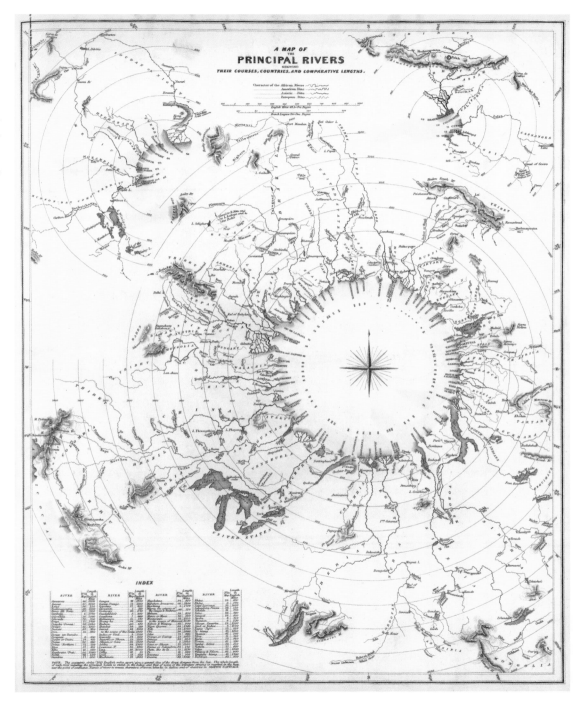

7

The Eclectic Archipelago and Lakes of All Countries

How strange to see them all together—the world's biggest islands and lakes. Or some of them, at least. For the selection criteria for inclusion in these weird collections seem to be based on the available space: The biggest islands and lakes are not included, and some relatively small ones are. The idea behind this eclectic collection seems to be to give students a sense of the comparative size of these normally very disparate bodies of land and water—although one is tempted to fit the islands into the lakes . . .

None of the ten largest islands managed to squeeze onto the list. Certainly not Australia, which is either the world's largest island or the smallest continent—but is usually excluded from the island list. The eleven largest undisputed islands are Greenland, New Guinea, Borneo, Madagascar, Baffin Island, Sumatra, Honshu (Japan's main island), Great Britain, Victoria Island, Ellesmere Island and Sulawesi.

The larger islands on the map are numbered according to their size (which is mentioned in square kilometers and square miles),

the others (smaller than 2,500 square kilometers, a bit less than 1,000 square miles) are not listed according to size.

12. Middle Island, New Zealand (145,836; 56,308)
14. North Island, New Zealand (111,583; 43,082)
16. Newfoundland, Canada (108,860; 42,031)
17. Cuba (105,806; 40,852)
18. Iceland (101,826; 39,315)
34. Southampton Island, Canada (41,214; 15,913)
43. Vancouver Island, Canada (31,285; 12,079)
53. New Caledonia, France (16,648; 6,467)
71. Jamaica (11,190; 4,320)
77. Cape Breton Island, Canada (10,311; 3,981)
80. Kodiak Island, United States (9,293; 3,588)
82. Puerto Rico, United States (9,100; 3,435)

85. Disko Island (now also known as Qeqertarsuaq), Greenland (8,612; 3,312)
87. Chiloé, Chili (8,478; 3,273)
90. Anticosti Island, Canada (7,941; 3,066)
99. East Falkland Island (6,605; 2,550)
113. Lawrence Island, United States (5,135; 1,983)
116. Trinidad, Trinidad and Tobago (5,009; 1,864)
117. Caviana Island, Brazil (5,000; 1,930)
124. West Falkland Island (4,531; 1,750)

Many of the other smaller islands are Caribbean: Mariguana, Eleuthera, Abaco, Bahama, San Salvador and Andros (all in the Bahamas); Trinidad and Tobago (which together form one nation); Grenada, Dominica, Saint Lucia (each an island nation), Saint Vincent (main island of Saint Vincent and the Grenadines); Anguilla (a British Overseas Territory); Guadeloupe (*sic*) and Martinique (each a French Overseas Department); and Santo Domingo (the island name actually is Hispaniola, which is divided between the

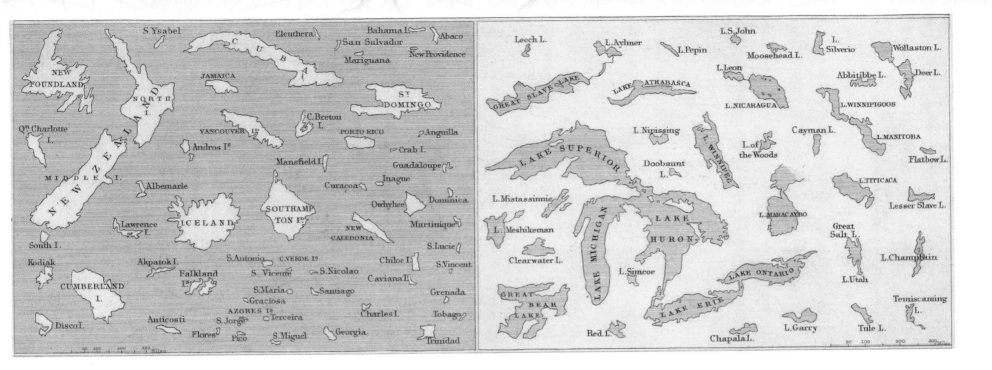

states of Haiti and the Dominican Republic).

Also figured are the individual isles of the Azores and Cape Verde archipelagoes (respectively Portuguese and formerly Portuguese possessions in the Atlantic Ocean).

Other islands are Akpatok (Canadian); Santa Isabel (the largest of the Solomon Islands in the Pacific); Hawaii (the Big Island of the eponymous state, mentioned here in the old spelling as Owhyhee); and South Georgia (mentioned here as Georgia, an uninhabited British island not far from the Falklands).

I was unable to identify Mansfield, Crab and Cumberland Islands. Charles and Albemarle Islands might be the Galápagos islands of Floreana and Isabela, respectively. Queen Charlotte Island, in Canada, is more correctly identified as plural, being composed of sev-

eral islands. New Zealand's "South" Island is called Stewart Island today.

Since maps are two-dimensional, it will be more useful to list the lakes mentioned here according to their surface, not their volume. Again, the list does not start with number one—in this case the Caspian Sea, in Central Asia. Lakes Michigan and Huron are counted as a single body of water because, well, they *are* a single body of water—connected via the Strait of Mackinac.

2. Lake Michigan-Huron, Canada/United States (117,702; 45,445)
3. Lake Superior, Canada/United States (82,414; 31,820)
7. Great Bear Lake, Canada (31,080; 12,000)
9. Great Slave Lake, Canada (28,930; 11,170)

10. Lake Erie, Canada/United States (25,719; 9,930)
11. Lake Winnipeg, Canada (23,553; 9,094)
12. Lake Ontario, Canada/United States (19,477; 7,520)
17. Lake Maracaibo, Venezuela (13,300; 5,100)
20. Lake Titicaca, Bolivia/Peru (8,135; 3,141)
21. Lake Nicaragua, Nicaragua (8,001; 3,089)
22. Lake Athabasca, Canada (7,920; 3,060)
37. Lake Manitoba, Canada (4,706; 2,924)
38. Great Salt Lake, United States (4,662; 2,897)

All the other lakes, practically all of them in the United States or Canada, are smaller than 4,500 square kilometers (about 2,800 square miles).

8

Some Like It Hot—and Wet: The Luscious Waterworld of Dubia

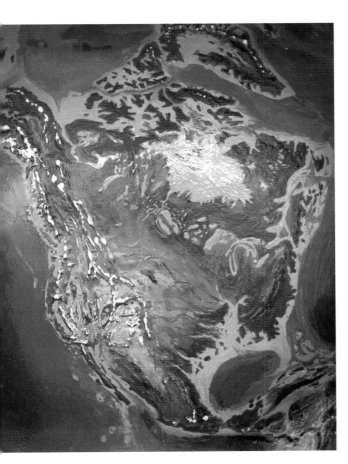

Welcome to planet Dubia, a strangely familiar place. It's what our planet could look like if you add a millennium's worth of global warming. By AD 3000, then, the equatorial regions are barely habitable, and warmer climes are the norm in formerly temperate zones. The polar ice caps have long since melted, raising sea levels by as much as 110 meters.

Coral seas encrust former coastal metropolises; Siberia is prime farmland while the prairies are now impenetrable jungles. "What if they don't change it back?" asks Chris Wayan, the artist who created Dubia, of that future world's inhabitants. "After all, they may argue, why put the Earth through birth-pains twice? Double jeopardy! It's climate *change*, not climate, that disrupts communities."

Mr. Wayan may have a point. For even though Dubia counts 30 million square kilometers (11.6 million square miles) of lowlands less than our planet, it's still a more fertile place—huge parts of Africa, Australia and Siberia are now arable. The earthlings of AD 3000 will probably find their climate perfectly natural. They might *like* it that hot—and consider us their unlucky Ice Age forebears.

Europe is a true continent at last, separated from Asia by the Ob Sea connecting the Arctic with the Caspian, Black and Mediterranean seas. Actually, Europe is more of an *archipelago:* Scandinavia is an island, as are Spain, Brittany and Normandy. An extended Baltic Sea harbors fragmented islands of what was mainland Europe. The Danube has become a giant estuary, cutting the Balkan Peninsula in half. Britain and Ireland are collections of smaller islands. London, Brussels, Amsterdam, Helsinki and St. Petersburg are gone, but on the upside, Paris, Vienna and Warsaw are now coastal cities. Farming will move north, but fishing will go south—the Mediterranean teems with life, thanks to its connections north and south, via the Suez Strait to the Indian Ocean.

In North America, a fertile Alaska now bristles with redwood forests, California boasts a Central Sea where the Valley used

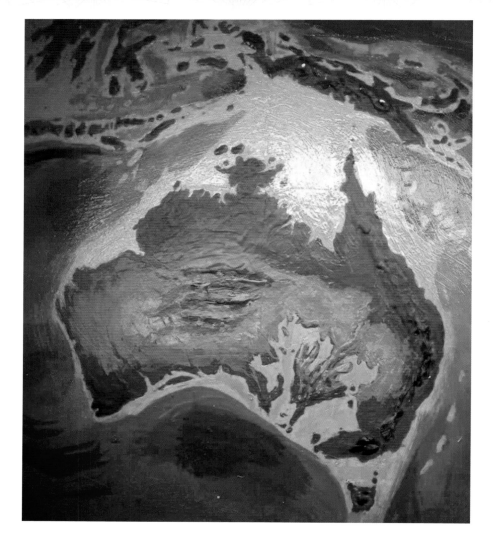

to be and, farther inland, the Bonneville Sea, back from prehistory, is only one of many in the greener, lusher West, which now looks more like East Africa at the dawn of man. The prairies have extended all the way up to Canada's heavily settled Arctic coast. Greenland has thawed to resemble a question mark, with a sheltered inland sea that sometimes still freezes over—the last remnant of our Ice Age—but flora and fauna thrive on the island

that finally lives up to its name. New England is an island, the big cities from Toronto to Washington drowned by the rising seas. Florida is still a tourist hotspot—but only for scuba divers, as it's completely submerged now. No more Louisiana either, and the Mississippi has created a wide bay, cutting deep into the American heartland all the way up to Illinois.

Australia is no longer a red desert with a green lining. On Dubia, its climate has tipped

from dry to monsoon—also thanks to the sea, which has advanced 115 miles from the north, bringing the equatorial rains south and turning desert into savanna and grassland. The advance of the sea nearly split Australia in two, in the process creating a Mediterranean climate around the Eyre Sea and the Darling Gulf. Queensland's tropical forests have expanded south into New South Wales, and even Tasmania is surrounded by coral reefs now.

XII. ICONIC MANHATTAN

Is New York the center of the universe or merely capital of the world?
What's certain is that Manhattan's very shape is as iconic as its skyline.

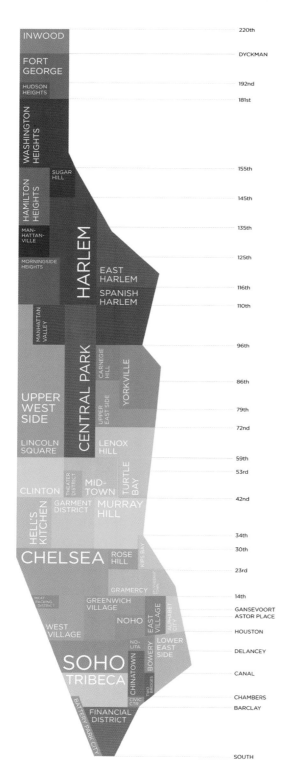

INWOOD	220th
	DYCKMAN
FORT GEORGE	
	192nd
HUDSON HEIGHTS	181st
WASHINGTON HEIGHTS	
	155th
SUGAR HILL	
	145th
HAMILTON HEIGHTS	
	135th
MAN-HATTAN-VILLE	
	125th
MORNINGSIDE HEIGHTS	
HARLEM / EAST HARLEM	116th
SPANISH HARLEM	110th
MANHATTAN VALLEY	
	96th
CENTRAL PARK / CARNEGIE HILL / YORKVILLE	86th
UPPER WEST SIDE / UPPER EAST SIDE	79th
	72nd
LINCOLN SQUARE / LENOX HILL	59th
	53rd
CLINTON / THEATER DISTRICT / MID-TOWN / TURTLE BAY	42nd
GARMENT DISTRICT / MURRAY HILL	
HELL'S KITCHEN	
	34th
	30th
CHELSEA / ROSE HILL / KIPS BAY	23rd
GRAMERCY	14th
MEAT PACKING DISTRICT / GREENWICH VILLAGE	GANSEVOORT ASTOR PLACE
WEST VILLAGE / NOHO / EAST VILLAGE / ALPHABET CITY	HOUSTON
NO-LITA / BOWERY / LOWER EAST SIDE	DELANCEY
SOHO / CHINATOWN / TWO BRIDGES	CANAL
TRIBECA / CIVIC CTR	CHAMBERS
	BARCLAY
FINANCIAL DISTRICT	
BATTERY PARK CITY	
	SOUTH

1

"No One Calls Clinton Clinton": Pinning Down Manhattan's Neighborhoods

Few geographic shapes are as instantly recognizable as as that of Manhattan Island, core of the Big Apple. The names of Manhattan's neigborhoods are just as iconic: SoHo, Little Italy, the Village, Hell's Kitchen, the Upper East Side, Harlem . . .

But the world-famous New York City neighborhoods' borders aren't as solid as those of Manhattan—neighborhoods shrink, move and expand, names change. This map offers a snapshot of Manhattan's constantly evolving subdivisions as might be perceived right about now.

This is a very debatable snapshot, as shown by these comments, received on the blog:

- "I live in 'Lenox Hill' and no one, as far as I know, says they live in 'Lenox Hill.' We say we live in the Upper East Side."
- "Find me a New Yorker who actually uses Nolita and I'll give you a bridge."
- "I can't believe Little Italy doesn't even register as a neighborhood anymore."
- "Where's Roosevelt Island?"

- "Pretty, but not too practical—Turtle Bay starts east of what avenue? Sugar Hill is between what avenues?"
- "It's interesting that they have a NoLiTa ('North of Little Italy'), but they don't have Little Italy. . . . It's still there, although Chinatown has certainly been extending itself northwards."
- "Glad to see a map that has both Hell's Kitchen and Clinton. Most have either one or the other."
- "I used to live near the tippy top of 'Clinton,' but nobody called it that. It was Hell's Kitchen. In fact, there was a campaign in the early nineties against 'Clinton'—they sold T-shirts and everything! Their beef was that real estate interests were trying to revive the 'Clinton' name to 'de-Hell' the place and make it sound marketable."
- "Hudson Heights does not exist. That is still Washington Heights."
- "Oh yeah, and Alphabet City? What is this, the 1980s or the Playbill for *Rent*?"

- "I come from Yorkville born and raised, and it went to Fifth Ave."
- "I live in Midtown, nobody has called it 'Radio City' for quite a while, but you could write me a letter in Radio City, NYC, and it'd reach me."
- "I agree there are neigborhoods missing: Yorkville, Knickerbocker. . . . And no one calls Clinton Clinton."
- "Rose Hill? There is very little hill, and no roses."
- "Murray Hill is smaller than shown: it does not extend below 34th street (even though the hill itself does), and it is bounded on the west by Park Avenue, on the east by Third. Nor does Murray Hill abut the Garment District: Midtown extends down between them."
- "Nix NOHO and Lincoln Square: these are in the East Village and the Upper West Side, respectively."

- "Tribeca, the 'TRIangle BElow CAnal street' is, in fact, triangular."
- "Not only is it attractive, but it's more accurate than the maps in the taxis."
- "I've been a tourguide since 1994. Never heard of 'Rose Hill' or 'Two Bridges' in that time."
- "Hell's Kitchen/Clinton are not separate neighborhoods but refer to same area (upper 30s to 59 west of 8th ave)."
- "The taxi maps show that some of Turtle Bay/Lenox Hill is considered Sutton-Beekman."
- "Yorkville used to be a more distinct neighborhood, but most of its German inhabitants have moved or died."
- "Very, very few people living in 'Fort George' consider it as such. Normal human beings call it Washington Heights."
- "This map, like most maps, forgets the northernmost neighborhood of Manhattan: Marble Hill. It is in what is now the Bronx; it was once a part of the island proper but a civil engineering project diverted the Harlem River south, cutting its ties to the island."
- "What I've noticed is that 'nice' neighborhoods—i.e., where pale people with decent incomes buy apartments (I'm one of them)—tend to get unique identities, but 'bad' neighborhoods, where poor people of color rent (and which, honestly, bear a striking resemblance to most of Manhattan in the 1970s), aren't allowed to keep their identities."

One more gripe about the shape of Manhattan itself: the island is actually longer than indicated on this map. As on many maps, its northern "tail" is severely truncated.

2

From Spuyten Duyvil to Battery Park: A Wordmap of Manhattan

There is much to wax lyrical about in the bustling borough of Manhattan, and indeed, this industrious island that is the beating heart of New York City has inspired poetry from Whitman, Ginsberg and many others.

This poem trumps all other Manhattan-inspired verse in one aspect: The poem is not just *about* the island, it *is* the island. The map was made by Howard Horowitz—not coincidentally a craftsman in both the fields of poetry and geography. He calls this and other products of his combined crafts "wordmaps."

These "wordmaps" are cartographic images that double as narrative texts, with as close a correspondence between the text and the localities it describes: The poem not only starts *with*, but also *at* Spuyten Duyvil, at the very north of the island. It ends at Manhattan's southern tip, at/with Battery Park. The seven textual extrusions away from the island correspond with some of the tunnels and bridges that connect the island to the outside world.

In all, "Manhattan" the poem describes over one hundred places on Manhattan the island. But the wordmap also touches upon the history, geology and ecology of the place, enumerates famous New Yorkers like Fiorello LaGuardia as well as infamous visitors such as Fidel Castro, lists the consecutive waves of immigration, sums up much of the fun to be had and culture to be enjoyed in the Big Apple . . . and much, much more.

Horowitz received the idea for this map in a vision during the blizzard of 1996, then worked at it for eighteen months, chiseling the island of words into shape "out of the bedrock of language and imagination."

As such, the map is a snapshot of an ever-evolving island: The Trade Towers mentioned near the southern edge of the island are no more. Other, less stark examples of the inexorable march of time intersperse the wordmap—but you'd need to be a Manhattanite to spot them. Manhattan is hardly an example of nature at its most pristine, but it shares its *horror vacui*. The urban fabric of Manhattan is constantly reweaving itself, making this wordmap not just an interesting geopoetic hybrid, but also a fascinating document of its time. In a hundred years' time, how many of these lines will have to be rewritten to update "Manhattan"?

The island's tip was sliced by a ship canal that tamed the Spuyten Duyvil shoals, but severed Marble Hill from Inwood. Medieval tapestry unicorns grace the Cloisters; a flag-pole and stockade mark old Fort Tryon. Lofty crags overlook the broad Hudson River as bedrock & history anchor the Heights to the George Washington Bridge. Walk east toward the Bronx across High Bridge; gaze to the south from Sugar Hill, where trumpeters and tap dancers stepped up into the sun. Ages ago Iapetus (an older Atlantic Ocean) closed; the kiss with Africa heated a melting pot. Lava was injected in veins of rock and coagulated to form Palisade cliffs. The legacy of Algonquian life is hidden in our place names and our meals. The new-comers (first the Dutch, then English, African, Irish, German, Italian, Jewish, Chinese, Greek, Ukrainian, Armenian, Puerto Rican, Pakistani, Cuban, Dominican, Haitian, Filipino, and all), have shed blood in a thousand places, but millions live. Legends of Gotham: Father Knickerbocker, Boss Tweed, Emma Lazarus, Fiorello, the roar of the El, the blizzard of '47, Giants at the Polo Grounds. Offshore, barges ply swirling brown water near North River sewage pipes, as striped bass and shad swim up "the river that flows both ways": a tidal reach of the sea all the way up to Albany. Brownstone, bodega, ball court & bus stop: on warm nights in Harlem, noisy streets and quiet rooftops. Kids splash around a hydrant as lovers embrace on a Riverside Park bench and rush-hour traffic is stalled on the Triborough Bridge. Some uptown options: gospel choir on Sunday, sooty Grant's Tomb, hiphop the Apollo, ribs at Sylvia's, law at Columbia, mangos in El Barrio, peace garden in the Cathedral, rowboat on the Meer, pub-crawl the West Side, listen to poetry at the 92nd St. Y, nosh at Zabar's, spiral up the Guggenheim, tour Gracie Mansion. Songbirds alight in leafy woods as a turtle lays eggs near a pond in Central Park. Grand museums flank the green with dinosaur bones and Egyptian tombs. When it snows, we ramble out to Sheep Meadow & the Great Lawn; in sunshine, to Strawberry Fields, the Lake, & the Zoo. Buy hot dogs from pushcarts near Madison boutiques, or hear grand opera at the Met. Step down to the world of subways. (Take the A train, ride the Lexington line, or change at 59th Street for the IRT. Catch the F out to Queens.) Gneiss but full of schist, the bedrock sparkles with mica. It bears the weight of midtown: skyscrapers at Columbus Circle, Fifth Avenue, and Park Avenue. Attend concerts at Carnegie, ice skating shows at Rockefeller Center, Mass at St. Patrick's Cathedral. Our eyes are drawn up to a blue slice of sky as vertical walls enclose us. 100 gridlocked taxis honk at police blockades as Fidel speaks at the U.N. Revelers jam Times Square on New Year's Eve, to jostle and sing as the ball drops. Buses come in (the Lincoln Tunnel) to Port Authority, trains to Grand Central. The lion-flanked public library was once a reservoir; we love the Art Deco classic Chrysler spire. From Hell's Kitchen walk to Broadway, buy tickets for "Showboat" or "Cats"— hey, the Knicks won at the buzzer in the Garden! See Macy's float parade, then gape from atop the Empire State, where mighty Kong took a fall. Diamond jewelers join fur-clad window shoppers as herds of jaywalkers cross against the light in the Garment District. Graffiti-scrawled boards near the Flatiron Building enclose pits of unconsolidated sediment Consolidated Edison must dig. Workers repair Gramercy Park cables, reroute Chelsea steam pipes, plug a burst main flooding streets by Union Square. (Tap water flows down from the Catskills in deep tunnels; garbage is hauled to a landfill at Fresh Kills.) The riverfront was filled for barnacle-crusted piers, and Minetta Brook wetlands became lots in Greenwich Village. A sweatshop horror: 146 locked-in women lost their lives in the Triangle Shirtwaist fire. Watch skateboard demons cavort among panhandlers as old men play chess near the arch in Washington Square. N.Y.U. students, art film fans, coffee drinkers, & East Village poets crowd smoky joints on Saturday night; some cross (the Holland Tunnel) back out to New Jersey. Cheap gallery space is a memory in SoHo; cast-iron lofts rent high, as do TriBeCa warehouses. A bag lady seeks warmth huddled over a sidewalk grate on the Bowery, where Stuyvesant's farm once spread in old New Amsterdam. The original steal (this island, traded for $24 in beads) lies plastered in myth and concrete, obscured like the African Burial Grounds. A Lower East Side delicatessen sells good chicken soup; enjoy zuppa di pesca at the Festival of San Gennaro, or bird's nest soup in Chinatown. Marchers to City Hall cross the Brooklyn Bridge to demonstrate, as tourists at South Street Seaport eat lunch with a view. The Fulton Fish Market is mobbed before dawn. Precambrian stocks bond the upper crust with solid foundations below the Trade Towers, Trinity Church and Wall Street. Ferryboats to Staten Island, the Statue of Liberty, Ellis Island, and Governor's Island depart from wind-swept docks at Battery Park.

3

A World Map of Manhattan

A history of successive waves of newcomers arriving in New York City, working their way up (or sideways) to make room for the next wave, arguably makes it the most emblematic immigrant city in the world.

This map celebrates that diversity by assembling Manhattan out of the contours of many of the world's countries. Danielle Hartman created the map based on data from the 2000 U.S. Census. In all, eighty different countries of origin were listed in the census. The mapmaker placed the country contours near the census area where most of the citizens of each country resided.

The title of this work is *New York—Global Island* to emphasize, in Hartman's words, "the relationship between Manhattan island and the final island design. The global island suggests that residents from all over the world can coexist, that they are integral to making the City what it is, and they can still retain their separate identities. Rather than a melting pot, the City is a rich mosaic, a microcosm of the world."

Vietnam is at the southern tip of Manhattan, joined there by a country that looks like Portugal and by Iraq, Italy and Spain, among others. China fills up the Lower East Side and, appropriately, Chinatown. Canadians and Australians seem to congregate mid-island, while Russians dominate the northern tip of Manhattan.

New York

Global Island

4

Who Put the Ice in NYC?

What would summer in New York be without ice cream?" Aaron Meshon painted Manhattan with his favorite icy treats in the picture. There's enough ice in the city during wintertime, which can be quite severe in New York. But summers can have a subtropical feel, with on average eighteen to twenty-five summer days with temperatures of 90° F (32° C) or more. What better excuse for an ice cream—whether in the shape of a cone, a lolly or a sandwich.

Meshon chose an eastward perspective for *N.Y.i.C.y*, painted as a wrap around a children's magazine, and prominently parked a giant ice-cream vendor near the Hudson River. On the far side of the island, the Queensboro Bridge is seen leaping over Roosevelt Island toward Queens. To the left, the square green patch of Central Park is visible. Right behind the vendor, the familiar shape of the Empire State Building looms. For those familiar with New York's skyscape, the rest of the painting presents the opportunity to "spot the building."

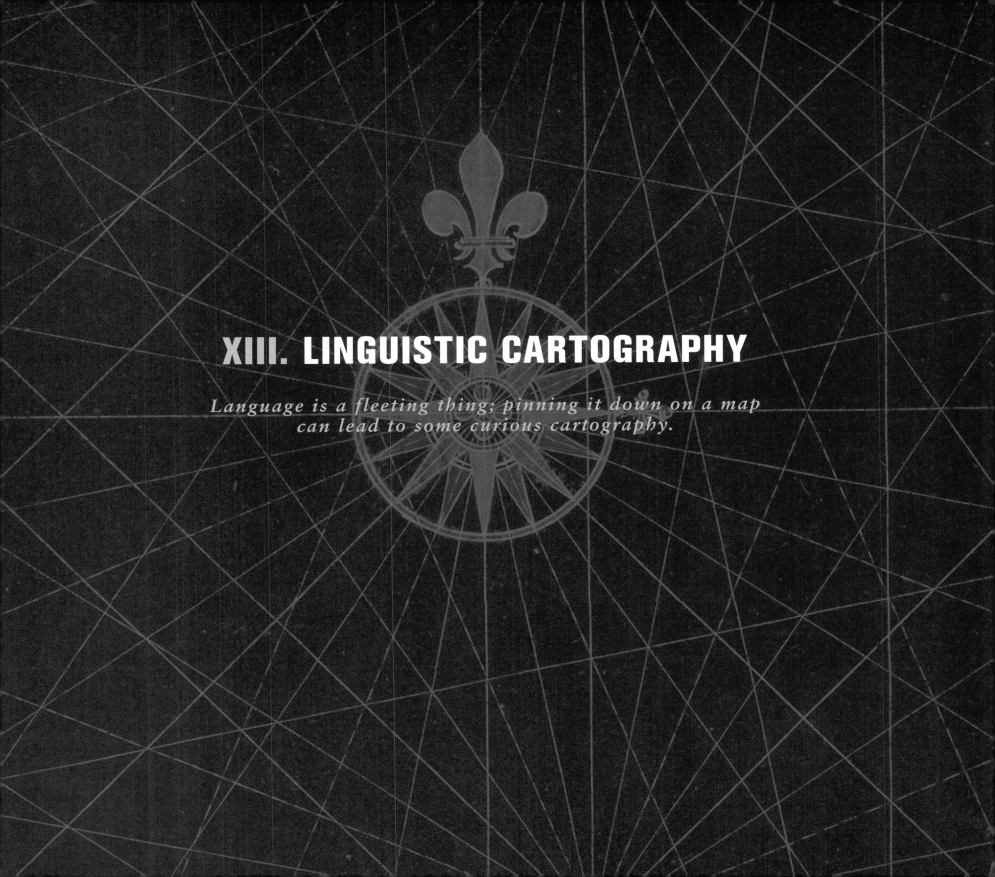

XIII. LINGUISTIC CARTOGRAPHY

*Language is a fleeting thing; pinning it down on a map
can lead to some curious cartography.*

1

The World's Linguistic Superpower:
Papua New Guinea

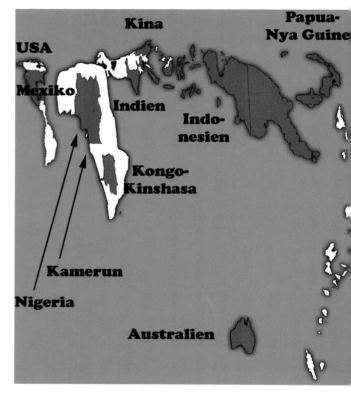

The book *Limits of Language* by the Swedish linguist Mikael Parkvall is a sort of languages-only *Guinness Book of World Records*, listing everything that's large, small and otherwise interesting about the manifold manners of human speech and associated forms of communication. One item deals with the world's most linguistically diverse countries, and is illustrated with this map, of the world's "linguistic superpowers." The caption reads:

Languages are very unevenly distributed among the countries of the world. The map tries to capture this fact by rendering each country in a size corresponding to the number of languages spoken in it. (Because of the inherent problems in accomplishing this, sizes are rather approximate). The ten shaded countries are those in which more than 200 languages are in use.

The *Ethnologue*'s most recent information lists the following ten countries as containing the largest number of living languages (indigenous and imported), corresponding with the countries on Mr. Parkvall's map:

1. Papua New Guinea: 820 languages
2. Indonesia: 742
3. Nigeria: 516
4. India: 427
5. United States: 311
6. Mexico: 297
7. Cameroon: 280
8. Australia: 275
9. China: 241
10. Democratic Republic of the Congo: 216

It's curious how the linguistically most diverse country in the world is Papua New Guinea (PNG)—because it's also the place with the biggest biodiversity anywhere, one of the last places in the world where new species get discovered regularly. Speculatively, this could be attributed to the extremely rugged terrain, isolating people, plants and animals in remote valleys, which in turn encourages the separate development of languages as well as species. The whole island of New Guinea (divided between the Indonesian province of Irian in the western half and the independent PNG in the eastern half) accounts for one in six of all languages spoken on earth.

One interesting PNG language that is somewhat intelligible to westerners (anglophones at least) is Tok Pisin, literally "Talk Pidgin"—based on simplified English. Tok Pisin is an official language, and the most commonly spoken one in PNG. Examples: "haus bilong king" (palace), "mi laikim yu tru" (I love you) and "win masin" (air conditioner).

2

Switzerland's Curious Culinary and Cultural Divide: The *Röstigraben*

Switzerland is predominantly German-speaking, but far from completely so. The alpine federation is officially quadrilingual: German (64 percent), French (20 percent), Italian (7 percent) and Romansh (0.5 percent). As the latter two languages are very minoritary, linguistic tension does tend to be a binary thing, between *Deutschschweiz*—a word only a germanophone could pronounce—and *la Romandie*, the francophone west of the country.

The Romands call the "other" side *la Suisse alémannique* and the *Schweizerdeutsche* call the francophone part of their country *Welschschweiz* (the root word being a Germanic term for "stranger," identical to the one in "Wales" and "Wallonia"). The language border dividing these two areas is known jestingly as the *Röstigraben* ("rösti ditch") in German and the *rideau de rösti* ("rösti curtain") or *barrière de rösti* ("rösti barrier") in French.

Not unlike hash browns or latkes, rösti is a dish made by frying grated potatoes in a pan. It was formerly eaten as breakfast by farmers in the (German-speaking) Bern canton. The original conceit of the *Röstigraben* was that it constituted the western limit of the German Swiss culture, beyond which people spoke (and ate) differently.

The rösti has gained popularity as a side dish all over Switzerland, but the language and cultural differences persist. The French Swiss voters have traditionally been less averse to the international community (including potential EU membership) and more likely to support a more active role for the federal government. Recently, voting trends in French and German Switzerland have tended to converge more.

The *Röstigraben* isn't the only gastronomically defined cultural border in central Europe. The northern and southern halves of Germany are separated by what is called the *Weisswurstäquator*—the white sausage equator, after a favorite dish in Bavaria that's rarely eaten in the north.

This map, used for the cover of a book about the phenomenon, shows a very literal *Röstigraben*—a Switzerland-shaped rösti broken in two exactly where the language border

runs. That the ditch wasn't too hard to cross is apparent by the name of the author, Laurent Flütsch: His French forename and German surname suggest his parents had a quite intimate knowledge of the "other" . . .

It has been suggested that this map is incomplete, or at least imprecise. The Italian-speaking southeast of Switzerland should really be made of polenta, the corn mush that is an essential part of the regional cuisine.

3

Paris's Most Unexpected Export: The Dorsal /r/

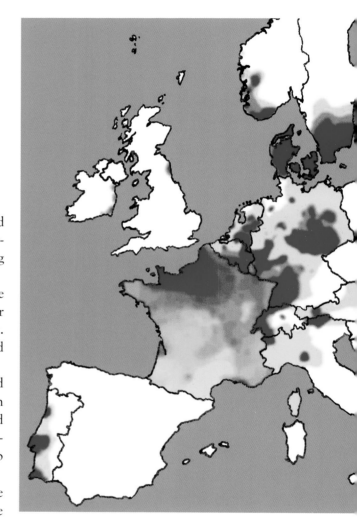

In the seventeenth and eighteenth centuries, something changed in the way Parisians spoke French. Increasingly, they started to pronounce the letter /r/ in a dorsal instead of an apical way. The apical /r/ is pronounced at the front of the mouth, using the tip of the tongue, while the dorsal /r/ is produced at the back of the throat, using the dorsum, or midbody, of the tongue.

"I would suggest that this is the most rapid spread by any linguistic feature, other than isolated lexical items, in documented history," says Mikael Parkvall.

For if the dorsal /r/ came from Paris, it must have been chic, and was therefore eagerly copied by elites elsewhere—and not just in the French language area. Even before the dorsal /r/ had spread to the rest of France, the feature emerged in Copenhagen, where it was first reported around 1780. Within a century and a half, the whole of Denmark was "conquered."

Sweden and Norway were affected by the Danish fashion, and the dorsal /r/ spread rapidly throughout mainland Scandinavia. Urban centers in Germany, Portugal and Italy picked up the fashionable pronunciation, and the dorsal /r/'s conquest continued overseas, reaching Québec, Brazil, New Caledonia, Israel . . .

A change this rapid requires people to have shifted from apical to dorsal /r/ during their lifetimes, as some research seems to suggest. In western Europe, only Spain and the United Kingdom remain marginally affected by it.

The dorsal /r/ nowadays is a standard feature of the French, German and Danish languages, optional in standard Dutch and Portuguese and regionally dominant in Italian, Swedish and Norwegian. This map shows the dorsal /r/ area in Europe today.

On this map, the darkest shades indicate where most people used the feature before the Second World War: northwestern France (including, of course, Paris), the south and east of Belgium and the Netherlands, central and southwestern Germany, all of Denmark and southern parts of Norway and Sweden, eastern Switzerland, the Tyrol areas of Italy and Austria and urban areas in Austria, Italy and Portugal.

The dorsal /r/ has since conquered all of France and most of the Low Countries, Germany, Austria and northern Italy. Intriguingly, according to this map, some coastal areas in the United Kingdom and Ireland are also affected by the fiendish dorsal /r/.

4

Praise the Lord and Pass the Dictionary: Europe's Polyglot Prayers

In 1741, Gottfried Hensel published *Synopsis Universae Philologiae* (A Summary of Universal Philology), containing this remarkable map of Europa Polyglotta (Multilingual Europe), using the opening lines of the Lord's Prayer as a kind of Rosetta Stone to show the diversity—and concordance—of some of the languages spoken in Europe.

It is quite probably the first linguistic map ever—and it leaves much to be improved upon. Europa Polyglotta was an atemporal rather than a contemporary language map of Europe, for it showed languages no longer spoken at the time (e.g., "Mauritanian" in the southern part of Spain). Nor was it complete, or very accurate, eighteenth-century linguistics still being very speculative. But fascinating nonetheless.

It shows alphabets used by several European peoples, such as (in the upper left corner, left to right): "the Scythians, born of the Hebrews," the Greeks, the Marcomanni, Runes, Moeso-Gothic, and Picto-Hibernic. In the upper right corner are shown *Characteri Rutenicae Lin-*

guae, i.e., the Russian alphabet. The lower left corner shows (left to right): the Latin, German and Anglo-Saxon alphabets. At the bottom, there are several other alphabets of the Hunnish, Slavonic (Cyrillic), Glagolitic (Illyric) and Etruscan (Eugubina) languages.

Many of the languages mentioned on the map itself are still widely in use, their version of the Lord's Prayer remaining quite understandable: Lusitania (Portuguese), Hispanica (Spanish), Biscainia sive Cantabrica (Basque), Catalani (Catalan), Gallica (French), Italica (Italian), Belgica (Dutch), Germanica (German), Dania (Danish), Noruegica (Norwegian), Suecica (Swedish), Finnonica (Finnish), Thulae Insulae (Icelandic), Lythuanica (Lithuanian), Polonica (Polish), Hungarica (Hungarian), Graeca (Greek).

Surprisingly, no contemporary English is used. The British Isles are covered by Hibernica (Irish Gaelic), Picto-Scotica (Scottish Gaelic) and Anglo-Saxonica (Old English). Other ancient rather than contemporary languages are Gothica (Gothic, apparently

spoken in the North Atlantic), Germanica Transmarina ("German from over the water," in central Scandinavia), Slavonica (Church Slavonic) and Tartaria ("a mixture of German and Slavic," spoken in the Crimea). Graeca Barbara (demotic Greek), placed in Asia Minor, is closer to modern Greek than the classical version placed in Greece.

Turkish is not located in present-day Turkey, but in Bulgaria (indeed still home to a sizable Turkish minority). Lapponica certainly isn't the Saami language, but apparently a northern variant of Finnish. Russia is inscribed with Tartarica (Tartar), Russica (Russian) and Nova-Zemblic—the most intriguing language of all, since Novaya Zemlya was uninhabited.

Some living languages absent from this map are Breton, Czech, Romanian, Bulgarian, Belarusian and Ukrainian. Other languages, dead or dying even then, are indicated: Ingerman, Livonian, Courlandish, Prussian. The tantalizing language labeled Foro-Juliania in southern France is probably Occitan, possibly

Friulian (which should then have been placed in northern Italy).

Hensel groups all those languages into three groups: Celtotheotisca ("Celtic-Germanic"), Illyrico-Slavonica ("Illyrian-Slavic") and Progenies Hellenica ("descendants of Greek," covering the former Roman Empire—i.e.,

the dark yellow—and roughly corresponding to the area of the Romance languages and Greek). English is aligned with the latter group, no doubt due to the large lexical influence of French on English.

Europa Polyglotta is one in a series of four maps, the others showing the language situa-

tion in the Americas, in Africa and in Asia. He appears to have based his work on Dan Brown's *Oratio Dominica: The Lord's Prayer in Above 100 Languages, Versions and Characters* (London, 1713).

5

Before France Spoke French

Latin once was the language of the Roman Empire, the Catholic Church and the Enlightenment. It's still taught in schools and used on some occasions by the church and the scientific community, but for all intents and purposes it's a dead parrot. There are no native Latin speakers. If Latin's dead, its successor languages haven't done at all bad. About 700 million people (over 10 percent of the world's population) speak a Romance language: French, Spanish, Portuguese, Italian, Romanian and the half dozen smaller languages derived from Latin. Only half as many people speak English as a first language.

Intriguingly, the spoken form of Latin, out of which the different Romance languages would evolve, started its conquest only after the fall of the empire. This map of France shows that the earliest latinization happened in the south (the Provence area) and small urban centers throughout the rest of the country, spreading out over large parts of the country in subsequent centuries, but leaving out large areas, even in the southwest, until the ninth century. Last latinized were France's extremities, where other languages were dominant (Dutch in the north, German in the northeast, Basque in the southwest, Breton in Brittany).

XIV. BASED ON THE UNDERGROUND

*With their colorful lines and abstract geography, the iconic maps
of the London Underground have inspired countless spin-offs.*

1

Oslo to Pyongyang Without Changing Trains: World Tube Map

You can call them Underground, Métro, Tube, U-Bahn or whatever—urban rail systems the world over offer a transport experience that is at once wildly divergent and weirdly similar. Now imagine combining all those systems. Imagine a global metro system, where you can get on in Newcastle, England, and get off in Esfahan, Iran, or Melbourne, Australia . . . without having to change trains!

This fantastic, daydream-inducing diagram was a promotional e-card for *Transit Maps of the World*, a coffee-table book by Mark Ovenden that between its covers assembles plans and layouts of the world's 230 or so urban rail systems (in use, construction or planning), which are linked together on this map. Produced in the style of Harry Beck's iconic 1933 London Underground map, it reveals the different degrees of metroization on different continents.

Africa is poorly endowed with public underground transit systems: Only Cairo and Alexandria (Egypt), Tunis (Tunisia), Algiers (Algeria) and Lagos (Nigeria) have or are planning them.

Actually, Oceania is even less metroized, but this is self-explanatory: There's no need for subways in a region where most countries are small island nations without large cities. Only Australia (Melbourne, Sydney) and New Zealand (Auckland)—significantly less small than the other Oceanic islands—have them.

Beck's method of making geography subservient to clarity distorts distances, in London as well as on this fanciful map—rendered even more bizarre by some unlikely stops close to each other: How about Baghdad to Izmir via Jerusalem, or Athens to Esfahan via Tel Aviv? Or Taipei to Pyongyang via Seoul?

As in Beck's design, there's a concentration of lines and stops in the central area (which on the London Tube map, I've only recently discovered, has the shape of a bottle). This gives the impression that outlying areas, such as the Americas, are much less metroized. Which might be a bit of an exaggeration, much like the placing of Bologna at the center of this world map is an overstatement, if ever so slightly, of that city's charm.

Okay, this is a fantasy transit map. But just imagine taking the metro in Vancouver, all the way to Shanghai! With stops in Montréal, Amsterdam, Prague, Kiev and Novosibirsk! Although that *is* a pretty long stretch to have to sit in a dark tunnel. Here are some of the comments sent in by visitors to the blog:

- "Cool . . . BUT, no direct route from Seattle to Vancouver, and no transpacific line?"
- "Nice map. Alas, the newest member to the club of cities with an urban rail transit system is missing from this otherwise interesting map. Please, welcome Charlotte (N.C.) to the ranks of cities with a rail-based transit system."
- "And Las Vegas and Los Angeles are in the wrong order—the highway from LA to SLC is divided by a stop in Las Vegas, in fact."
- "Definitely an attempt to stay true to

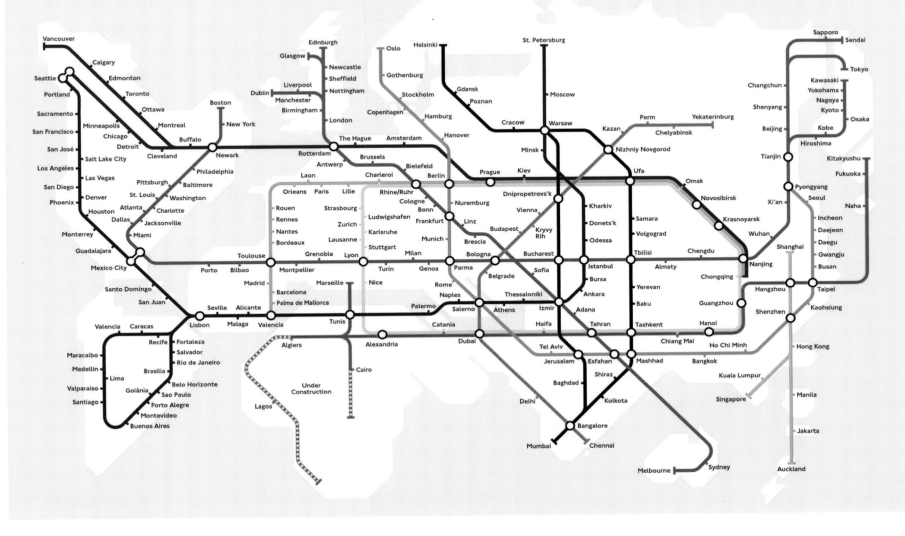

the original tube map, at the expense of topographic accuracy. In Europe there are all kinds of strange orderings (Gothenburg to Copenhagen via Stockholm? Amsterdam to Ruhrgebiet via Bielefeld? don't think so)."

- "Of course, there's a really good reason the 'London Underground' look works so well: the Tube is very North-London-heavy, and cities with metros are very Northern-Hemisphere-heavy."
- "I'd avoid the Tehran to Tel Aviv segment."
- "Since when is Vienna north-east of Budapest?"
- "Ha this is great! According to this map, I commute from Nizhny Novgorod to Catania everyday, that's so much more exciting than from Highbury and Islington to Sloane Square!"
- "Montevideo and Ottawa don't have a subway!"
- "Houston is more like a sorry excuse for a transportation system—lots of buses, and one light rail that nobody uses."
- "It's a shame that Taiwan had to lose its status as an island in order to keep the jubilee line running smoothly."

2

Itineraries into Eternity (and Back, Occasionally)

If the world's religions can be seen as itineraries through life, then they might be represented as a variation on Harry Beck's Tube map. Lines 1 and 2 (green) represent the Islamic religious path, with both lines calling at stations symbolizing finding the truth, embracing faith and conforming to God's will. Death is followed by the penalty of the grave, after which ensues the Last Judgment. Only then will the righteous be separated from the not-so-righteous; the latter will follow line 1 to Hell, where gruesome trials await them—scorpion stings, bitter herbs for food, ear-shattering shrieks of sorrow. The former will be able to travel via line 2 to Heaven, where they will sit on precious carpets, eat divine food and have every wish fulfilled.

Lines 3, 4 and 5 (purple) represent the Christian (or, more correctly, the Catholic) path. Line 3a, throwing unbaptized babies straight into Limbo, was "closed" by Pope Benedict XVI in 2006. All other lines lead

through baptism, respecting the Ten Commandments and confessing sins to death. Line 3 leads to Hell, but not before a stop at Purgatory, where one might have one's sins purged and still get to Heaven, via line 4a to the Resurrection and the Revival of the Dead. Line 4 skips Purgatory, straight to Resurrection, and eventually Heaven. Line 5 is an express lane, skipping the long dirt nap before Resurrection and heading straight for Heaven.

Line 6 (blue) is the Jewish religious path, where everybody quite straightforwardly sins, dies, gets resurrected and revived and gets to Heaven. The only specification is an express lane between sinning and death—those who sin hasten their own demise.

Somewhat more complicated is line 7 (red), the shamanic religious path. For shamans possess the ability to visit the Other Side while still alive (via the SR line, the Shamanic Voyage). Shamans can also detect when people are about to die; the animal

soul returns to earth as a ghost, the shadow soul joins the clouds and the sky, and a third part of the soul goes on a Death Voyage, in which the shaman can assist. The soul of the deceased can choose to appear as a ghost to the living (or in some versions of this belief system even reincarnate), or choose the Death of the Soul, into the Empire of the Dead. Again, some versions of shamanism connect certain tests and trials with this transition, such as having sex with a mystical being. Sometimes, Hell is imagined as a separate place beyond the Empire of the Dead.

Lines 8 and 9 (brown) denote the Hindu view of things, which is the circular system of Samsara, allowing personal enlightenment through one of four ways: devotion (bhakti yoga), action (karma yoga), knowledge (jnana yoga) or the "way of the king" (raja yoga). "Bliss" is the necessary state of mind to attain enlightenment, the ultimate goal of which is to avoid the circularity of reincarnation, and reach Moksha, the release from reincarnation.

Lines 10, 11 and 12 (orange) deal with the Buddhist way of life (and death). Its essence is simple: Life equals suffering, suffering is caused by desire, desire can be overcome by attaining Nirvana, Nirvana can be reached through an eightfold path: true realization, true attitude, true reasoning, true conduct, true life, true endeavor, true mindfulness, true contemplation.

After death, the karma of one's actions remains, and is the seed of one's next life. On the forty-ninth day after death, the soul is reborn, either as a human, a deva, a demon, a hunger spirit or an animal . . . Or as a creature of Hell.

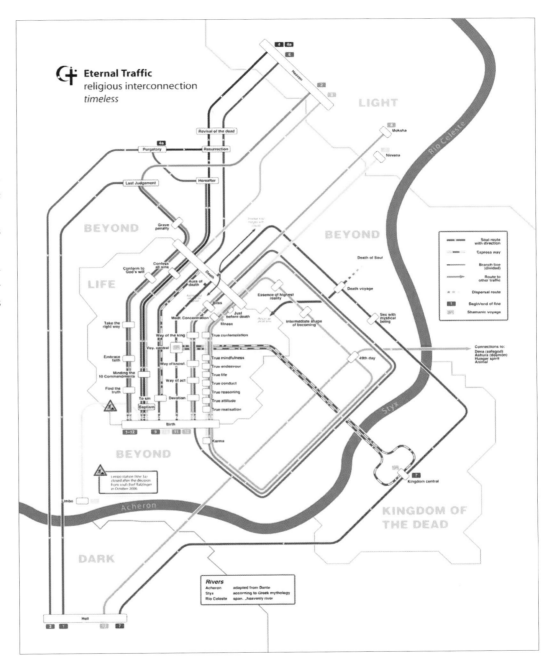

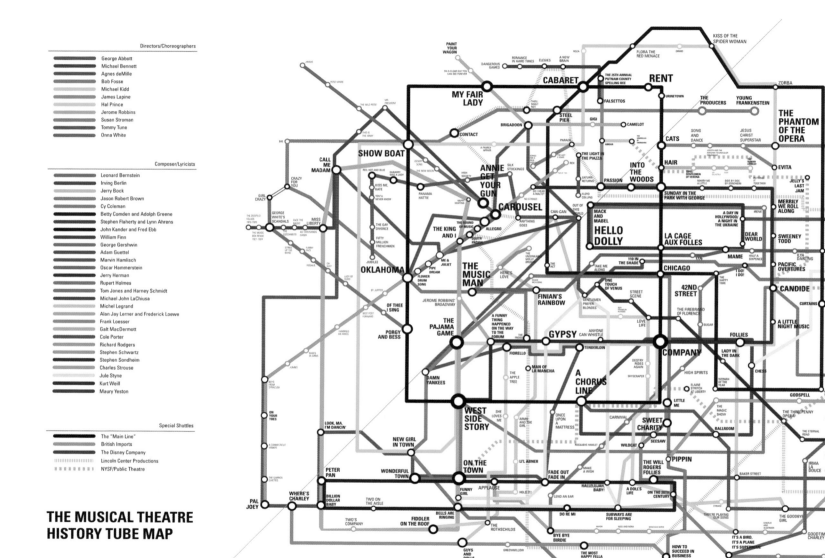

THE MUSICAL THEATRE
HISTORY TUBE MAP

CREATED BY **JOHN HOWREY**
LEARN MORE AT CHONNY.COM

Directors/Choreographers

- George Abbott
- Michael Bennett
- Agnes deMille
- Bob Fosse
- Michael Kidd
- James Lapine
- Hal Prince
- Jerome Robbins
- Susan Stroman
- Tommy Tune
- Onna White

Composer/Lyricists

- Leonard Bernstein
- Irving Berlin
- Jerry Bock
- Jason Robert Brown
- Cy Coleman
- Betty Comden and Adolph Greene
- Stephen Flaherty and Lynn Ahrens
- John Kander and Fred Ebb
- William Finn
- George Gershwin
- Adam Guettel
- Marvin Hamlisch
- Oscar Hammerstein
- Jerry Herman
- Rupert Holmes
- Tom Jones and Harney Schmidt
- Michael John LaChiusa
- Michel Legrand
- Alan Jay Lerner and Frederick Loewe
- Frank Loesser
- Galt MacDermott
- Cole Porter
- Richard Rodgers
- Stephen Schwartz
- Stephen Sondheim
- Charles Strouse
- Jule Styne
- Kurt Weill
- Maury Yeston

Special Shuttles

- The "Main Line"
- British Imports
- The Disney Company
- Lincoln Center Productions
- NYSF/Public Theatre

3

Cecil B.'s Niece and Other Musical Bee's Knees

Like some primeval forest, the history of American musical theater is both wide and dense and easy to get lost in. John Howrey produced a breadcrumb trail for his students by way of this Musical Theatre History Tube Map. Tube, yes, as in the London Underground, although the map traces the particulars of *American* musicals—but then there is some overlap between the shows on New York's Broadway and those in London's West End.

There are two groups of lines, the first representing some influential directors and choreographers, the second standing for famous composers and lyricists. The better-known names (at least for a non-aficionado like yours truly) tend to be in the second group: Bernstein, Berlin, Gershwin, Hammerstein, Porter, Sondheim, Weill. Just one name stands out in the first group: de Mille, but only because Agnes is Cecil B.'s niece. No idea whether she really was the bee's knees.

The stations represent the musicals on which those people worked or, as it happened, collaborated. "The Main Line was created to provide a main connection point and to make a more circular pattern rather than let the spider web run wild," says Howrey. "To make it, I chose two shows from each decade: one definite work, and one that pushed the boundaries."

Other special lines include a selection of British imports, musicals by the Disney Company and by the Lincoln Center. The size of the name of each particular musical on this map is related to its importance and the length of its run.

4

A Diagram of the Eisenhower Interstate System

The Dwight D. Eisenhower National System of Interstate and Defense Highways (Eisenhower Interstate System for short, or EIS for even shorter) spans the entire United States, including Alaska and Hawaii. At a total length of 46,837 miles (75,376 km), the EIS is the largest highway system in the world. It serves all major American cities.

This diagrammatical overview of the EIS manages to do what few maps of such a huge road network can—give an overview of the entire system that is both complete (detailing all of the over 220 interchanges) and convenient.

Some useful facts about the EIS:

- The EIS was initiated by Eisenhower in 1956, partly because he was impressed with the German system of autobahns, which provided easy transport also for military purposes.

- The system was considered complete as of September 15, 1991, when the last traffic signal was removed from I-90 in Wallace, Idaho.
- The accumulated cost of the EIS was $114 billion (original estimate: $25 billion).
- Highways running east-west are given even numbers, those running north-south are assigned odd numbers. As a general rule, odd-numbered highways increase from east to west, and even-numbered ones from south to north. And highways divisible by 5 generally are bigger and longer than others.
- Most interstates have two numbers (I-4, I-5 and I-8 are the only single-digit interstates outside of Hawaii, which has H-1, H-2 and H-3).
- Three-digit interstates are auxiliary highways. In principle, if they connect two interstates, the last two digits will be of a primary interstate, preceded by an even number; if they only connect at one end to a primary interstate, the first number will be odd.
- A widespread urban legend states that one out of every five miles of the interstate highway system must be built straight and flat so as to be usable by aircraft during times of war; this is not true.
- The most heavily traveled part of the interstate highway system is the 405 Freeway in Seal Beach, California, with a 2002 estimate of 377,000 vehicles a day. The least traveled section is I-95 just north of Houlton, Maine, with 1,880 vehicles a day (2001 estimate).

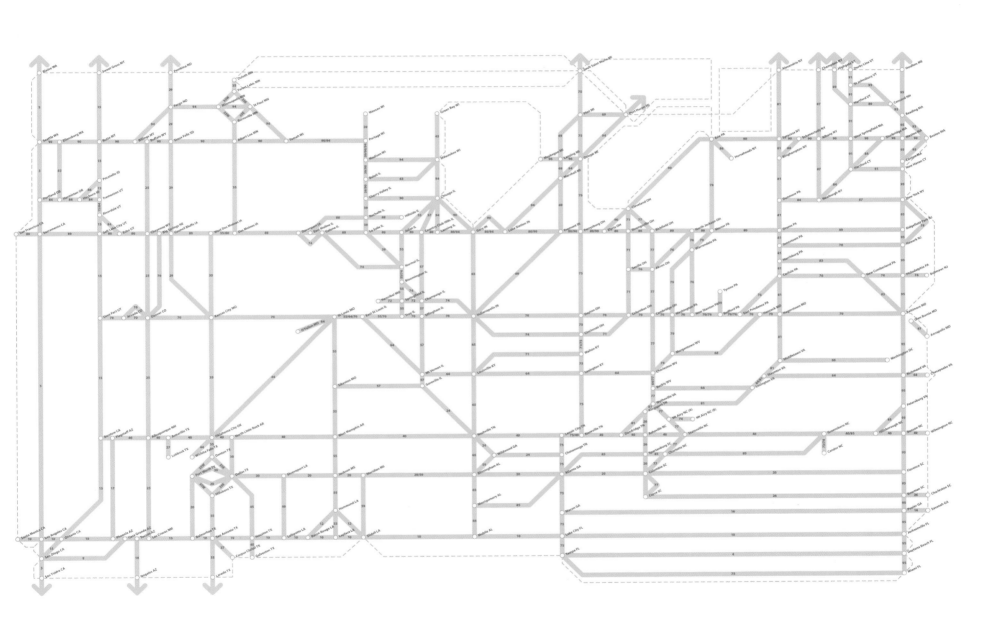

XV. FANTASTIC MAPS

Where imagination, fantasy and cartography meet.
Maps consisting of equal parts fantasy and cartography.

1

The Whole World in a Cloverleaf

Heinrich Bunting (1545–1606) knew the world didn't really look like this. There are enough maps in his works (including the *Itinerarium Sacrae Scripturae*) to indicate he knew the continents had an irregular and not a symbolic shape.

Yet he delighted in drawing other symbolic maps, examples of which can be anthropomorphic (Europe as a virgin) or hippomorphic (Asia as a winged horse). This particular map is a tribute to Bunting's hometown, Hanover, as the text above the map indicates: *Die ganze Welt in einem Kleberblatt welches ist der Stadt Hannover meines lieben Vaterlandes Wapen* ("The whole world in a cloverleaf, which is the coat of arms of Hanover, my dear Fatherland").

The map shows a world divided into three parts (Europe, Asia and Africa), connected at a single central point: Jerusalem. This is essentially still the same symbolic map of the world as the one first devised by Saint Isidore in the seventh century. Isidore's "T and O"–shaped map, itself inspired by scripture, influenced Christian European mapmaking up until the Age of Discovery.

That age would be the one Bunting grew up in. He and his contemporaries were among the first generations of Europeans to know Isidore was wrong—but it's almost impossible to resist imagining how this centuries-old archetype of a map took a while to be erased from the common memory of cartographers.

Bunting's map is nice in that it combines symbolism with realism: In the bottom left corner a piece of land is named America. Strange is that a similarly detached piece of territory at the top of the map is labeled Denmark and Sweden. Bunting must have known that Denmark was contiguous with the European continent. And England, floating in the ocean above France, looks suspiciously like a cap as worn by soldiers in the American Civil War. A red patch in the water between Africa and Asia is labeled as the Red Sea.

Some named countries and places (not all are easily readable) on the three continents are, left to right:

- Europe: Hispanien (Spain), Mailand (Milan), Welschland (an early German name for Italy), Frankreich (France), Lothringen (Lorraine), Roma (Rome), Deutschland (Germany), Behemen (Bohemia), Ungarn (Hungary), Polen (Poland), Reussen (Russia), Griechenland (Greece), Türken (Turks), Saaren (might be Saarland), Moscharo (Moscow?)
- Africa: Lybia, Egypten, Morenland (Land of the Moors), Königreich Melinde (Kingdom of Melinde), Caput Bonae Spes (Cape of Good Hope); three cities are named: Alexandria, Cyrene, Meroë.
- Asia: Siria, Arabia, Mesopotamia, Armenia, Chaldea, Persia, India; many cities are named, among them: Damascus, Ur, Babylon, Susa, Persepolis, Antioch, Nineveh.

Die gantze Welt in einem Kleberblat/Welches ist der Stadt Hannouer meines lieben Vaterlandes Wapen.

Dennemarck
Schwe-den
SEPTENTRIO
Engeland
Franckreich
Saxen
Hispanien
Deudschland
Lothringen
Behemen
Reussen
Meiland
Polen
EVROPA
Vngern
Moschaw
Welschland
Türcken
Griechen-land
Roma 382.
ARMENIA
MEDEN
Niniue 171.
Rages 349.
MESOPO TAMIA
PER SIA
INDIA
ASIA
SIRIA
Haran 110.
Persepolis
CHALDEA
Antiochia 70.
Babylon 170.
Susa 230.
Damascus 40.
Vr 156.
ARABIA
Gaza 312.
IERVSALEM
Das Rote Meer.
Das grosse Mittelmeer der Welt.
Alexandria 72.
Egypten
Cyrene 204.
LYBIA
Meroe 24.
Morenland
AFRICA
Köngreich Melinde.
CAPVT BONÆ SPEI
MERICA
ie Newe
Welt.

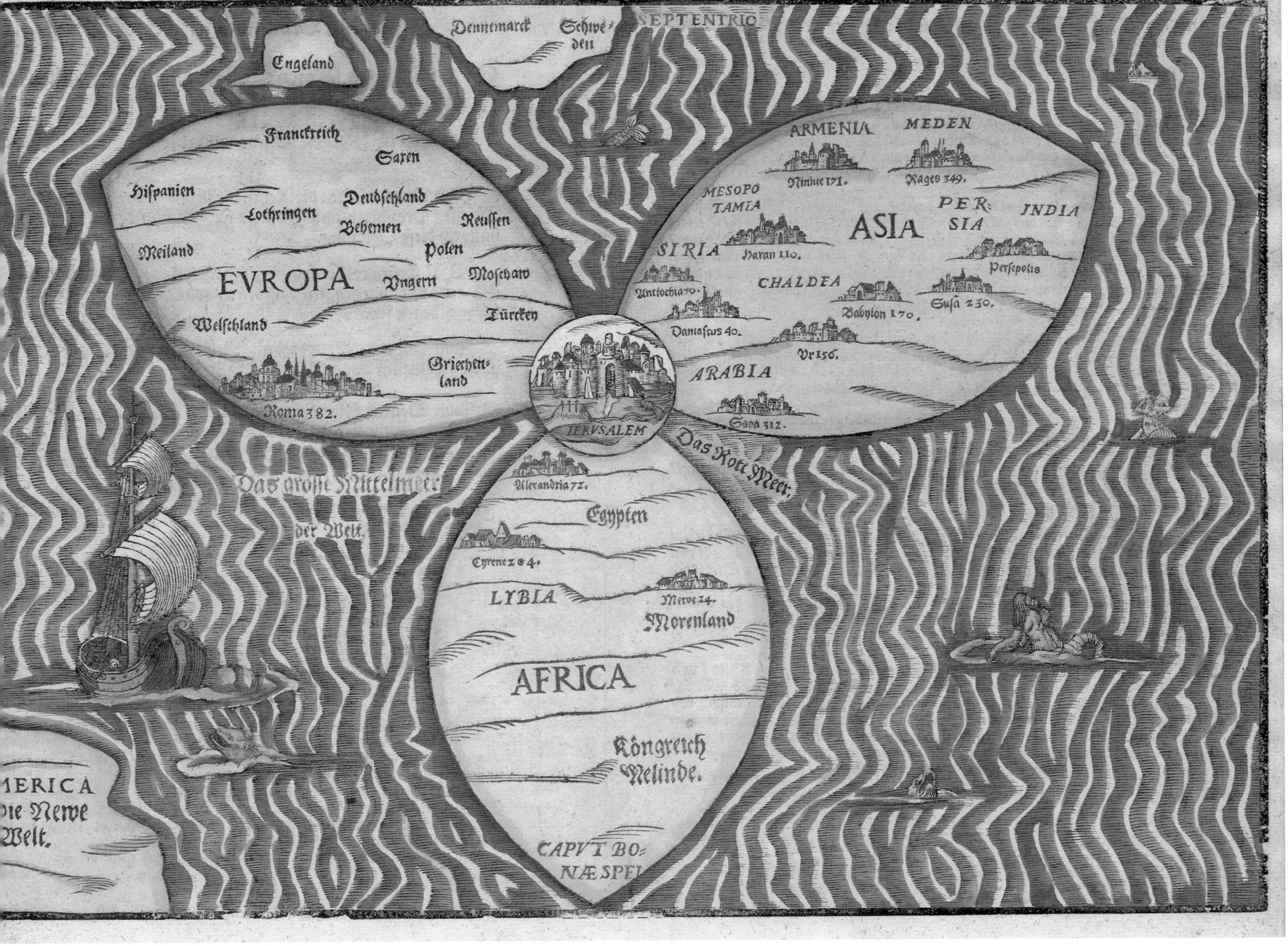

2

Jamerica the Beautiful

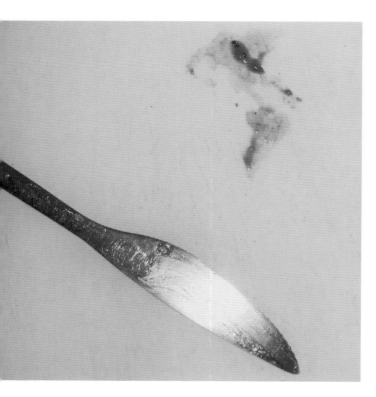

People of a very religious disposition have been known to see the face of Jesus in a slice of burnt toast, or the Virgin Mary's silhouette in a tree. Map nuts similarly observe pareidolia of states and continents in everyday objects.

"I've seen photos of clouds resembling maps," writes Bjørn A. Bojesen. "But never a blob of jam." But then: "I was just making a sandwich, and there it was—America on the chopping board!"

Both the Atlantic and Pacific coastlines of South America are wonderfully rendered, the accretion of jam at the left-hand side even symbolizing the elevation of the Andes mountain chain. Central and North America are somewhat less true to life, but their general shape is not that far off. As jam-based maps go, anyway.

However, "there is something really weird going on in Alaska," as Mr. Bojesen readily admits. The Aleutian Islands have morphed from a narrow island chain into a gigantic terrestrial tentacle, sticking into the Pacific Ocean and almost touching the West Coast. On the other side of the continent, Cuba and/or other Caribbean islands have hypertrophied and are drifting east into the Atlantic.

Some commentators on the blog misread the name, and thought it should read Jamaica. "Jamerican is what a Jamaican fellow in my office calls himself," said one. Another thought this might explain why Americans so often vote conserve-ative.

3

One Ring to Rule Them All, Mate: Tolkien's Australia

The koality of muh-cy is not strined": I forget who once pondered the impossibility of believing Shakespeare spoken in an Australian accent. Maybe it's the implied anachronism, for in Shakespeare's time there wasn't an Australian accent, owing mainly to Australia not having been discovered yet.

At first glance this map, transposing Tolkien's fantasy world on Australia, seems equally out of place. The imagined continent of Middle-Earth has always been taken to represent or at least prefigure Europe. The Hobbits, for example, are, says Tolkien, "just rustic English people, made small in size because it reflects the generally small reach of their imagination."

And yet putting the eurocentrism of fantasy cartography to one side, it's worth recalling that the *Lord of the Rings* movie trilogy wasn't filmed in Hungary, England or anywhere else in Europe, but in New Zealand—Australia's neighbor—ironically, about as far away in the world from Tolkien's rustic English country folk as you can get.

So if New Zealand can be the (rather spectacular) backdrop to Tolkien's stories, why not Australia? Disbelief duly suspended, let's examine the places mentioned in this map, mainly existing Aussie names that have been tolkienified:

- Western Australia ("Westron Australia")
- Perth ("Middle-Perth")
- Broome ("Brun")
- Alice Springs ("Alfalas Springs")
- Lake Eyre ("Lake Corseyre")
- Hobart ("Hobartton"—a nice reference to Hobbiton)
- Sydney ("Sidnarin")
- Queensland ("Quenyasland")
- Adelaide ("Adeleade")
- Brisbane ("Brohan")
- Melbourne ("Morborn")
- Northern Territory ("The Northern Waste")
- Great Dividing Range ("Great Dividing Rangers")
- Canberra, the federal capital, isn't too popular with the mapmaker ("Here was of old the witch-realm of Canbrar")

Incidentally, there *is* a place in Australia that is named after one in Tolkien's Middle-Earth: Mordor Pound. Previously named Spring Pound, this geographic feature in Australia's Red Center, about 40 miles northeast of Alice Springs, was renamed for Lord Sauron's dark realm in the 1970s, when geologist Alan Langworthy was struck by the resemblance between the two landscapes, one imaginary and one real.

This map was made by James Hutchings, an Australian himself. "A great place for a holiday," he says about his tolkienified Australia, "but watch out for the kangarorcs."

4

Oh, Inverted World: If Land Was Sea and Sea Was Land

As we've all learned in school, 70 percent of the earth's surface is covered by water, only 30 percent is solid ground. What if everything was reversed? What if every landmass was a body of water, and vice versa?

This map explores that question, and it is fantastic in at least three definitions of that word: fanciful, implausible and marvelous. The interior of China is marked by a spouting whale, a sailboat plows the waves of the Brazilian Ocean, a school of fish traverses the watery wastes of Siberia, large cities dominate places rarely frequented by people in this universe . . .

The oceans in this inverted world are the Great Eurasian Ocean (the world's largest) and the African, North and South American and Antarctica oceans. These are punctuated by islands that in our world are lakes:

- Baikal Island: surely a mountainous place, as in our world it is the deepest, most voluminous freshwater lake on the planet, containing 20 percent of the world's liquid fresh surface water.

- To the west of this vast ocean, close to the Mediterranean landmass, lies the unnamed Caspian Island—to the east thereof is the tiny (and if the reversal is symmetrical, rapidly sinking . . . or should that be emerging? Can't work that one out) Aral Island.

- A similar island, unnamed in the map, perforates Africa; this must be Lake Victoria, or rather Victoria Island.

- Other similar landmasses are the Great Islands, substituting for the Great Lakes. In this map, perhaps intentionally, they look like a dolphin doing a show jump.

- The seas are merely dotted with boats and fish, but 70 percent of the planet's surface is now walkable, arable, livable, mappable.

- The Gulfstream Mountains form the backbone of the North Atlantic States (I'm not sure whether the Eifel Tower close to the African shore is part of them).

- The South Atlantic Kingdom is marked by giraffes galloping near the Brazilian shore, a cactus and a burning sun. An important population center is St. Helena City, close to our world's British island dependency of Saint Helena.

- The narrows between the Brazilian and Antarctica oceans are dominated by Drake City.

- At the western shores of the South American Ocean lies the South Pacific Kingdom.

- To the north, on the eastern shore, is Mexico Land (in our Gulf of Mexico).

- Bermuda City seems to form a separate entity from the North Atlantic States.

- Labrador City lies between the United Ocean and the Greenland Sea (shouldn't that be frozen?).

- Celtic Land lies just to the south of English Lake.
- Surely, the Mediterranean Kingdom is a pivotal player, located between the western outlet of the Great Eurasian Ocean and the African Ocean—with a land bridge extending toward the Indian Kingdom.
- Flocks of sheep and windmills dominate the vast expanses of land toward the Australian Sea.
- The Philippine Kingdom rules the Far East, punctuated by the Japan, Taiwan and Philippine lakes.
- On the far northern shores of the Great Eurasian Ocean, finally, lie Arctic Land and the East Siberian Kingdom.

XVI. CARTOGRAMS AND OTHER DATA MAPS

*Maps can be used to display more types of data
than mere geographical information.*

1

Turn the Other Cheek: The French Kissing Map

At the risk of disappointing a few readers: This is not a map of French kissing, but a French map of kissing. For kissing in France is a much more complex subject than the country's somewhat overstated reputation for carefree libidinosity implies.

Unlike more reserved nationalities, the French greet each other with kisses on the cheek: men and women, women and women, and sometimes—if they know each other well enough—also men and men. Don't try this at home!

Unfortunately, the practice varies to the point where one risks *l'embarras social* if the kisser has another number of pecks in mind than the kissee. Imagine that you intend to give three kisses and the other person turns away after only two. *Ah, quel humiliation!*

Combien de bises? (How many kisses?) To answer this conundrum once and for all, Frenchman Gilles Debunne set up a Web site, asking his fellow countrymen and -women to send in how many kisses were de rigueur in their particular *département*. Over 30,000 votes have been cast; the reported number of kisses, two in most places, but varying from one to five, shows an interesting regional distribution.

- One kiss is the preferred option in only two *départements*: Finistère at the western tip of Brittany and Deux-Sèvres in the Poitou-Charentes region. Intriguingly, Deux-Sèvres borders three different kiss areas: the four-kiss zone in the north, the two-kiss zone on its eastern and western sides, and a single three-kiss *département* (Charente) to the south.
- Apart from Charente, all ten other three-kiss *départements* form a contiguous area in France's southeast. *Trois bises* are the thing to do in Ardèche, Aveyron, Cantal, Drôme, Haute-Loire, Hautes-Alpes, Hérault, Gard, Lozère

and Vaucluse. Hemmed in by Italy, the Mediterranean and the three-kiss area, four *départements* insist on only two kisses.
- Four kisses are the norm in a broad band of twenty-two *départements* in northeastern France, stretching from the English Channel and the Bay of Biscay to the Belgian border in the north (isolating one- and two-kiss areas in the west and north).
- A notable exception is a two-kiss enclave, made up of Paris (the four *départements* enlarged at the left-hand side) and the adjacent *départements* of Yvelines and Essonnes.
- The rest of the country sticks to just two pecks on the cheek, except—of course—the island of Corsica, known for its fiery temperaments (Napoleon was born there). Corsicans are the only French to kiss each other *five* times.

Mr. Debunne's map does not tell all, however. First off, there's the confusion *within départements*. Moving your mouse cursor over any *département* will show that none speaks with one voice; in the Ardennes, the northernmost four-kiss *département*, those four kisses are preferred by a plurality of only 42 percent, with about 30 percent preferring just two, and around 10 percent each opting for one, three or five kisses.

And then there's the dynamic aspect of this map. If you check out Mr. Debunne's Web site, some *départements* might have switched to another preference—although the large contiguous areas tend not to shift dramatically.

Keeping all that in mind, plus the fact that distinctions of age, class, occasion and familiarity may codetermine the number of kisses, it's probably wise to shake hands when you meet someone for the first time, and inquire about the desired number of kisses when you part with a new acquaintance. Or be creative, like this reader: "As three kisses are the rule here when celebrating, congratulate the other party even if there is no direct cause, so at least they'll know how many kisses to expect!"

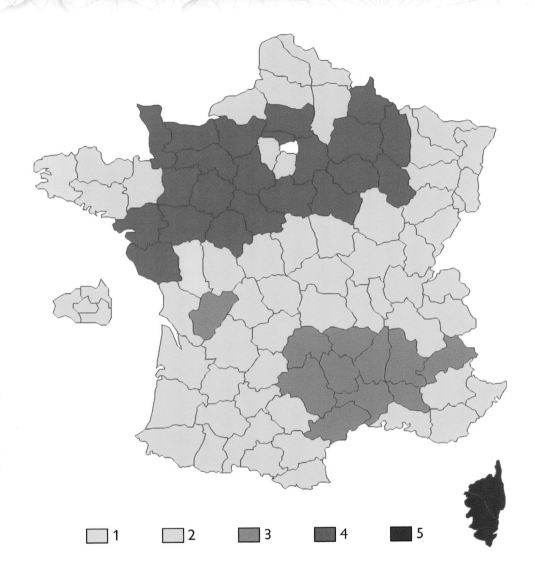

☐ 1   ☐ 2   ☐ 3   ☐ 4   ☐ 5

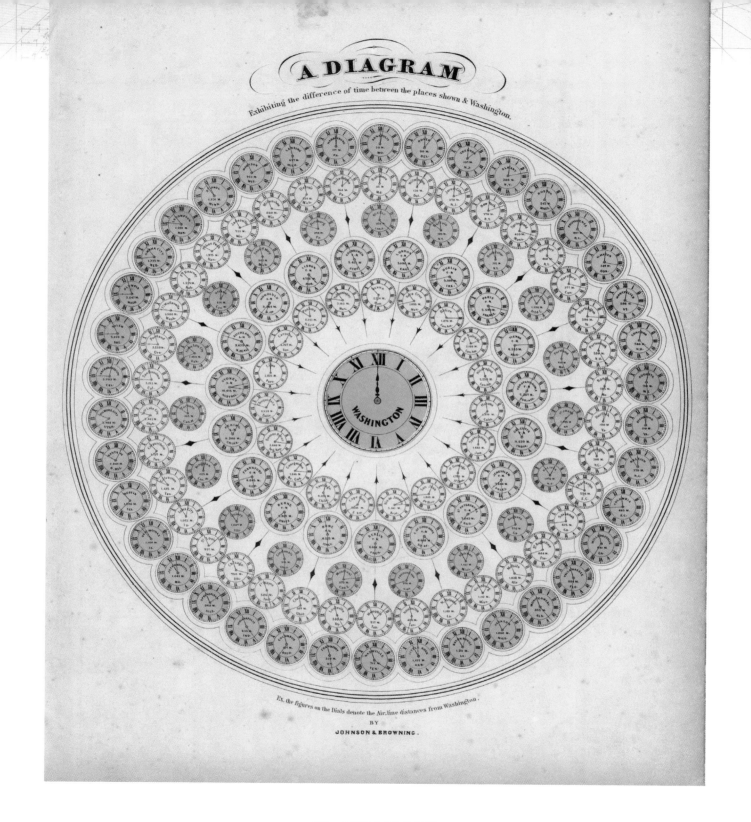

2

High Noon in Washington, D.C.: Keeping Time Before Trains

No Hillary, it's not 3 a.m. in Washington—it's noon, as indicated by the dial at the center of this diagram. Surrounding it are 133 similar dials, each of different cities throughout the United States and the world, each with its own time.

This diagram, "exhibiting the difference of time between the places shown & Washington," dates from 1860—and back then, this was just about the handiest way of knowing what the local time was in, say, Baltimore or Chicago.

For this was still twenty-three years before the introduction of standard time zones in North America. Every town had its own local mean time—it was noon when the sun was at its highest above that particular locality. Noon in Washington, D.C., meant two minutes past noon in Baltimore, although that city's only thirty-five miles (60 km) to the northwest of

D.C. At the same time, it'd be 12:12 in New York City, but still only 11:43 in Savannah.

As in Europe, it was the growth of the rail network that made all those local times extremely impractical. Just try to imagine having to draw up a train schedule, and you'll understand why on October 11, 1883, the railroad bosses met in Chicago to adopt the Standard Time System, swiftly adopted by most states (and rather less swiftly by the federal government, which took nearly half a century to officialize the practice).

This diagram gave pre–standard time people an overview of 133 local mean times, which only worked perfectly, I should think, if you were in D.C. and it was noon or midnight.

The inner (yellow) circle details nineteen localities in Latin America, from Mexico (10:32) through Cape Horn (12:40) to Lima (11:59).

The second (green) circle indicates the time in nineteen places in Europe, Asia and Oceania, such as London (17:08), Jerusalem (19:29) and Sydney (03:13).

The third (red) circle gives the time for nineteen cities in eastern Canada and the eastern United States, for example Ottawa (12:05), Portland, Maine (12:27), and Sioux Falls City in the Dakota Territory (10:42).

The fourth (yellow) and fifth (red) circles show the time for thirty-eight places each throughout the rest of the United States, from San Francisco (08:58) to Saint Augustine (11:42).

Inside each dial is a number denoting "the Air-line distance from Washington." The word "airline" has acquired another, more dominant meaning since 1860; of course, we today would call this a beeline.

3

Vital Statistics of a Deadly Campaign: The Minard Map

The best statistical graphic ever drawn" is how statistician Edward Tufte described this chart in his authoritative work *The Visual Display of Quantitative Information*.

The chart, or statistical graphic, is also a map. And a strange one at that. It depicts the advance into (1812) and retreat from (1813) Russia by Napoleon's Grande Armée, which was decimated by a combination of the Russian winter, the Russian army and its scorched-earth tactics—the retreating Russians burned anything that might feed or shelter the French, thereby severely weakening Napoleon's army.

As a statistical chart, the map unites six different sets of data:

- Geography: Rivers, cities and battles are named and placed according to their occurrence on a regular map.
- The army's course: The path's flow follows the way in and out that Napoleon took.
- The army's direction: This is indicated by

the color of the path, gold leading into Russia, black leading out of it.

- The number of soldiers remaining: The path gets successively narrower, a plain reminder of the campaign's human toll, as each millimeter represents 10,000 men.
- Temperature: The freezing cold of the Russian winter on the return trip is indicated at the bottom, in the republican measurement of degrees of réaumur (water freezes at 0 degrees réaumur, boils at 80 degrees réaumur).
- Time: This is given in relation to the temperature indicated at the bottom, from right to left, starting October 24 (*pluie*, i.e., rain) to December 7 (−26 degrees).

Pause a moment to ponder the horrific human cost represented by this map: Napoleon entered Russia with 442,000 men, took Moscow with only 100,000 men remaining, wandered around its abandoned ruins for some time and escaped the east's wintry clutches

with barely 10,000 shivering soldiers. Those include 6,000 rejoining the "bulk" of the army from up north. Napoleon never recovered from this blow, and would be decisively beaten at Waterloo less than two years later.

Almost exactly a century and three decades later, Hitler would repeat Napoleon's mistake by again underestimating the vastness of Russia, the inhospitability of its winters and the determination of the Russians.

The *Economist*, which in its last issue of 2007 published a story on the way in which some charts succesfully visualize statistical data (yes, those editorial meetings must be a riot), pointed out that "as men tried, and mostly failed to cross the Berezina river under heavy attack, the width of the black line halves: another 20,000 or so gone. The French now use the expression *C'est la Bérézina* to describe a total disaster."

The map was the work of Charles Joseph Minard (1781–1870), a French civil engineer who was an inspector-general of bridges and roads, but whose most remembered legacy is

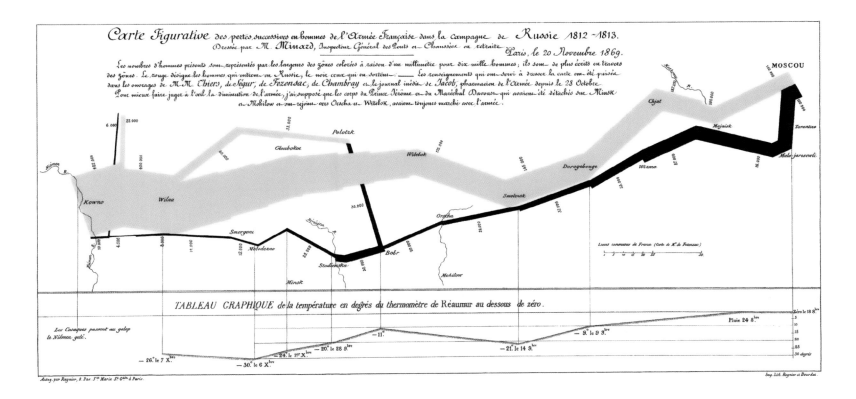

in the field of statistical graphics, producing this and other maps in his retirement. This is a translation of the legend at the top of the map:

Figurative chart of the successive losses in men by the French army in the Russian campaign 1812–1813. Drawn up by Mr. Minard, inspector-general of bridges and roads (retired). Paris, November 20, 1869.

The number of men present is symbolized by the broadness of the colored zones at a rate of one millimeter for ten thousand men; furthermore, those numbers are written across the zones. The red (*sic*) signifies the men who entered Russia, the black those who got out of it.

The data used to draw up this chart were found in the works of Messrs. Thiers, de Ségur, de Fezensac, de

Chambray and the unpublished journal of Jacob, pharmacist of the French army since 28 October. To better represent the diminution of the army, I've pretended that the army corps of Prince Jerôme and of Marshall Davousz which were detached at Minsk and Mobilow and rejoined the main force at Orscha and Witebsk, had always marched together with the army.

4

If Wikipedia Was China, Then . . .

In terms of Wikipedia, are you Chinese, Macedonian, Montenegrin or Grenadan? Frank San Miguel, self-declared software geek (and a veteran of Internet start-ups such as MapQuest) took some Wikipedia contributor statistics, added a few maps and came up with this nifty little piece of statistical cartography.

"Compare the population of world countries to the Wikipedia contributors. In the hierarchy of users, the vast majority of visitors to Wikipedia, 217 million of them, are readers; for the most part, they don't edit articles. Next are the regular contributors, who have contributed more than 10 times ever. There are about 340,000 of those. Next are the 105,000 active editors, who contribute between 5 and 100 times per month. Finally, there are the 10,000 very active editors who contribute more than 100 times per month.

"So if Wikipedia readers are like China, then the contributors are like Macedonia (regular), Montenegro (active) and Grenada (very active)."

Wikipedia Contributor Map

Population	Wikipedia Users
China, 1.3B	217M Readers (<10 edits total)
Macedonia, 2M	340K Regular Contributors (>10 edits)
Montenegro, 690k	105k Active Editors (>5 edits/mo)
Grenada, 90k	14k Very Active Editors (>100 edits/mo)

5

Ulstermen, Tennesseans, Mexicans and Germans:
A Tale of Texas's Squares

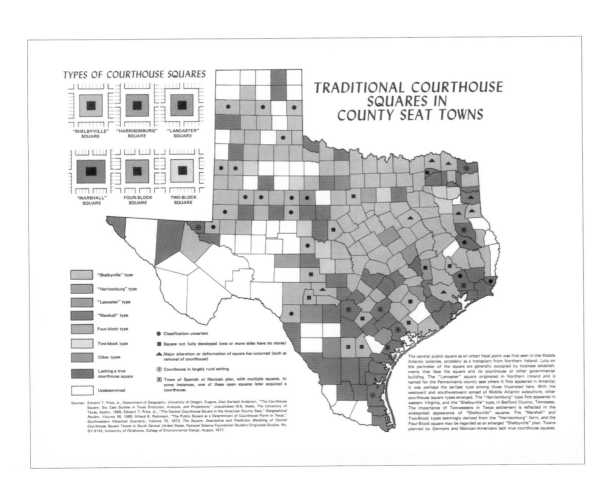

TYPES OF COURTHOUSE SQUARES

"SHELBYVILLE" SQUARE   "HARRISONBURG" SQUARE   "LANCASTER" SQUARE

"MARSHALL" SQUARE   FOUR-BLOCK SQUARE   TWO-BLOCK SQUARE

TRADITIONAL COURTHOUSE SQUARES IN COUNTY SEAT TOWNS

"Shelbyville" type

"Harrisonburg" type

"Lancaster" type

"Marshall" type

Four-block type

Two-block type

Other types

Lacking a true courthouse square

Undetermined

● Classification uncertain

■ Square not fully developed (one or more sides have no stores)

▲ Major alteration or deformation of square has occurred (such as removal of courthouse)

⊙ Courthouse in largely rural setting

▣ Town of Spanish or Mexican plan, with multiple squares. In some instances, one of these open squares later acquired a courthouse.

The central public square as an urban focal point was first seen in the Middle Atlantic colonies, probably as a transplant from Northern Ireland. Lots on the perimeter of the square are generally occupied by business establishments that face the square and its courthouse or other governmental building. The "Lancaster" square originated in Northern Ireland and is named for the Pennsylvania county seat where it first appeared in America; it was perhaps the earliest type among those illustrated here. With the westward and southwestward spread of Middle Atlantic subculture, other courthouse square types emerged. The "Harrisonburg" type first appeared in western Virginia, and the "Shelbyville" type, in Bedford County, Tennessee. The importance of Tennesseans in Texas settlement is reflected in the widespread appearance of "Shelbyville" squares. The "Marshall" and Two-Block types seemingly derived from the "Harrisonburg" form, and the Four-Block square may be regarded as an enlarged "Shelbyville" plan. Towns planned by Germans and Mexican-Americans lack true courthouse squares.

Sources: Edward T. Price, Jr., Department of Geography, University of Oregon, Eugene. Also Garland Anderson, "The Courthouse Square: Six Case Studies in Texas Evolution, Analysis, and Projections," unpublished M.S. thesis, The University of Texas, Austin, 1968; Edward T. Price, Jr., "The Central Courthouse Square in the American County Seat," *Geographical Review*, Volume 58, 1968; Willard B. Robinson, "The Public Square as a Determinant of Courthouse Form in Texas," *Southwestern Historical Quarterly*, Volume 75, 1972; *The Square: Descriptive and Predictive Modeling of Central Courthouse Square Towns in South Central United States*, National Science Foundation Student Originated Studies, No. GY-9142, University of Oklahoma, College of Environmental Design, August, 1971.

One of Northern Ireland's many under-appreciated contributions to the United States is the central public square as an element of urban planning. This architectural feature, centered around a courthouse, first appeared in the Middle Atlantic colonies as the "Lancaster" type square, so named after the Pennsylvania county whence it spread.

In a typical Lancaster square, the courthouse is surrounded by four L-shaped buildings, covering each corner and opening up into streets leading up to the middle of each side of the central square.

Modifications soon occurred: the "Harrisonburg" type was developed in western Virginia, the "Shelbyville" type in Tennessee. The "Marshall" and "Two-Block" types sprang from the Harrisonburg type, while the "Four-Block" square may be seen as a refinement of the Shelbyville groundplan.

Only one Texas county originally had a Marshall-type courthouse square; it's unclear how much remains, as the legend to the symbol in the county reads, "Major alteration or deformation of square has occurred (such as removal of courthouse)." Only two have the original Lancaster courthouse square layout (but neither in a "perfect" form).

The Four-Block courthouse square is almost as rare, occurring in only three counties, followed by the Harrisonburg square (four counties) and the Two-Block square (nine counties).

The bulk of classifiable courthouse squares in Texas belong to the Shelbyville type. This is indicative of the important role played by Tennesseans in the settlement of Texas in the early part of the nineteenth century. Judging by where the Shelbyville type was im-plemented, these Tennesseans had massive, widespread influence throughout the whole of Texas; only in the far north, the southwest and the far south is this type of courthouse square totally absent. This is partly due to the fact that Germans and Mexicans did not con-struct their towns around central courthouse squares.

6

Super Interesting: The World from a Brazilian Perspective

OS PAÍSES DENTRO DO BRASIL

A economia do Brasil é a 9ª maior do mundo – tão grande que os estados brasileiros têm o mesmo PIB (a soma das riquezas geradas anualmente num país) que países inteiros. São Paulo é tão rico quanto toda a Argentina. No Nordeste, a riqueza dos estados equivale à de países da África e da América Central. Pernambuco é tão rico quanto o Camboja, Sergipe equivale à Jamaica. Dá pra fazer a mesma comparação com os estados americanos (veja a seguir).

FONTES: IBGE (Contas Regionais do Brasil 2004 – PIB a preços correntes, segundo as Grandes Regiões e Unidades da Federação), Banco Mundial (PIB por paridade de poder de compra e população mundial 2006) e Prefeitura de São Paulo.

OS PAÍSES DENTRO DE SÃO PAULO

Este é o mapa de São Paulo, 4ª cidade mais populosa do mundo. Só a cidade, sem contar a região metropolitana, tem a mesma população que 5 países. A zona sul, com 3,5 milhões de habitantes, tem tanta gente quanto o Uruguai; a zona norte, com 2,2 milhões, tem a população do Kuwait. Também existe um Timor Leste dentro da cidade: na zona oeste, que tem 970 mil habitantes.

A VERDADE NOS MAPAS

Trocamos o nome dos lugares nos mapas para mostrar o verdadeiro tamanho da economia, da população e da violência ao redor do mundo.

TEXTO STEFAN GAN E LEANDRO NARLOCH   ILUSTRAÇÕES MARTINI

Inspired by the Strange Maps blog, Brazilian magazine *Superinteresante* presented a series of data maps comparing violence in Europe and Brazil, the economies of Africa to the size of multinationals and the strength of the GDPs of U.S. and Brazilian states to those of other countries.

A. We Are the World—Now Also in Brazil

Brazil, now the world's ninth largest economy, is a huge country with one foot in the developing world and the other in the developed. The economy of its states is equivalent to the gross national product of entire countries— albeit usually quite poor ones or small ones. This is especially the case out west, where the states are big, sparsely populated and underdeveloped, e.g., the three biggest states: Amazonas (Zimbabwe), Pará (Paraguay) and Mato Grosso (Bolivia). The more substantial economies seem to be located at the coast, and particularly in the south: Rio Grande do Sul (United Arab Emirates), Paraná (New Zealand),

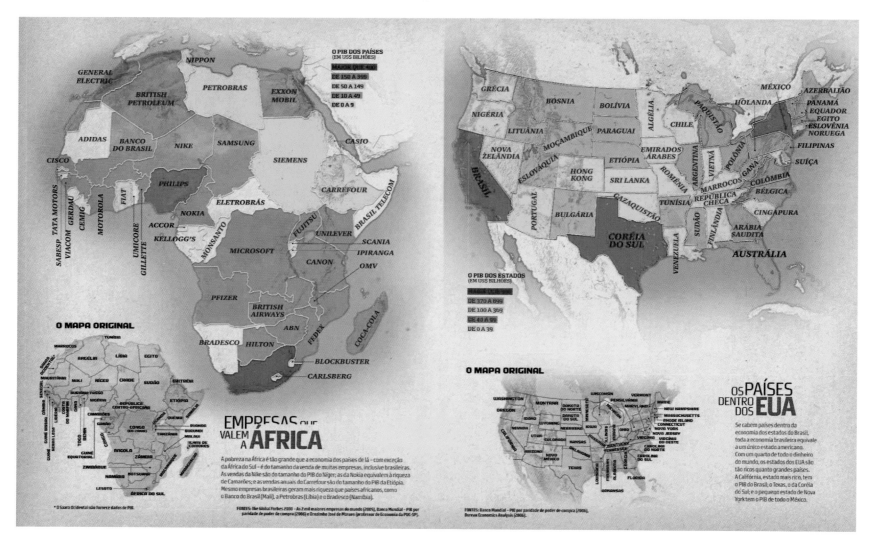

O PIB DOS PAÍSES
(EM US$ BILHÕES)

MAIOR QUE 400
DE 150 A 399
DE 50 A 149
DE 10 A 49
DE 0 A 9

O PIB DOS ESTADOS
(EM US$ BILHÕES)

MAIOR QUE 1.036 M.
DE 370 A 899
DE 100 A 369
DE 40 A 99
DE 0 A 39

O MAPA ORIGINAL

EMPRESAS QUE VALEM A ÁFRICA

A pobreza na África é tão grande que a economia dos países de lá - com exceção da África do Sul - é do tamanho da venda de muitas empresas, inclusive brasileiras. As vendas da Nike são do tamanho do PIB do Níger; as da Nokia equivalem à riqueza de Camarões; e as vendas anuais do Carrefour são do tamanho do PIB da Etiópia. Mesmo empresas brasileiras geram mais riqueza que países africanos, como o Banco do Brasil (Mali), a Petrobras (Líbia) e o Bradesco (Namíbia).

*O Saara Ocidental não fornece dados de PIB.

FONTES: The Global Forbes 2000 - As 2 mil maiores empresas do mundo (2005), Banco Mundial - PIB por paridade de poder de compra (2006) e Drauzinho José de Moraes (professor de Economia da PUC-SP).

O MAPA ORIGINAL

OS PAÍSES DENTRO DOS EUA

Se cabem países dentro da economia dos estados do Brasil, toda a economia brasileira equivale a um único estado americano. Com um quarto de todo o dinheiro do mundo, os estados dos EUA são tão ricos quanto grandes países. A Califórnia, estado mais rico, tem o PIB do Brasil; o Texas, o da Coréia do Sul; e o pequeno estado de Nova York tem o PIB de todo o México.

FONTES: Banco Mundial - PIB por paridade de poder de compra (2006), Bureau Economics Analysis (2006).

Argentina (São Paulo), Rio de Janeiro (Denmark), Bahia (Kuwait)!

B. Companies Are Equivalent to African Countries

African poverty is still so great that the economy of the continent's countries can be equated with the annual sales figures of some multinational companies. Sales for Nike are equiva-lent to the gross national product of Niger, Nokia sells as much in a year as the economy of Cameroon produces. Carrefour and Ethiopia are equivalent. Carlsberg is Lesotho, Micro-soft the Democratic Republic of the Congo, Coca-Cola is Madagascar. Even a regional heavyweight such as Nigeria still is no bigger than electronics multinational Philips. The one exception is South Africa, with an economy important enough to escape comparison with a single company; it is the only African coun-try with a GNP superior to $400 billion.

C. My State's Economy Can Beat Up Your Country's GDP

The U.S. economy, valued at $13.8 trillion in 2007, represents about a quarter of the entire world economy and still is the biggest single economy in the world. Nothing brings home its size like a comparison, on state a level, between the American economy and that of other countries around the world. California, the state with the largest economy within the United States, has a gross state product equivalent to the GDP of Brazil. Texas (the second largest economy within the United States) equates to South Korea, New York's economy (third largest) has the same size as Mexico's. The pairing of state economies to those of foreign countries leads to some incongruous duos: tiny Massachusetts and gigantic Egypt; sweltering Alabama and frigid Finland; ever-rainy Washington State and Mediterranean Greece.

D. Brazilian Cities Are as Violent as European Countries

Brazil is one of the most violent countries in the world; Europe is the most peaceful continent. Some Brazilian cities have about the same murder rates as European countries, despite having only a fraction of their population. In 2002, the city of Osasco (a suburb of São Paulo, counting 680,000 inhabitants) registered 506 murders, a figure comparable to Spain's 494 homicides, despite the fact that it has 45.2 million inhabitants.

Other comparable homicide rates: Curitiba (1.8 million) and Germany (82.2 million), Salvador (2.9 million) and England (50.8 million), and Recife (3.6 million) and France (64.5 million).

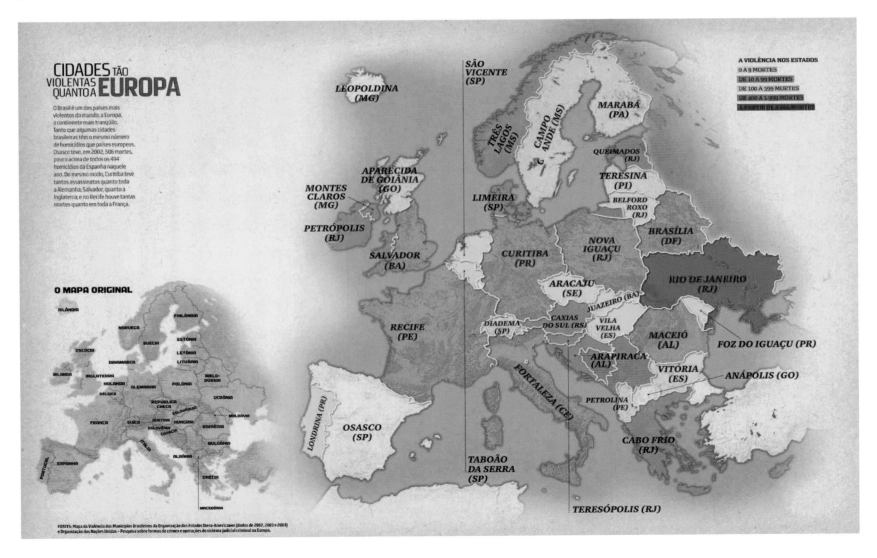

CIDADES TÃO VIOLENTAS QUANTO A EUROPA

O Brasil é um dos países mais violentos do mundo; a Europa, o continente mais tranqüilo. Tanto que algumas cidades brasileiras têm o mesmo número de homicídios que países europeus. Osasco teve, em 2002, 506 mortes, pouco acima de todos os 494 homicídios da Espanha naquele ano. Do mesmo modo, Curitiba teve tantos assassinatos quanto toda a Alemanha; Salvador, quanto a Inglaterra; e no Recife houve tantas mortes quanto em toda a França.

A VIOLÊNCIA NOS ESTADOS
0 A 9 MORTES
DE 10 A 99 MORTES
DE 100 A 399 MORTES
DE 400 A 1 999 MORTES
A PARTIR DE 2 000 MORTES

O MAPA ORIGINAL

FONTES: Mapa da Violência dos Municípios Brasileiros da Organização dos Estados Ibero-Americanos (dados de 2002, 2003 e 2004) e Organização das Nações Unidas – Pesquisa sobre formas de crimes e operações do sistema judicial criminal na Europa.

7

Where's Australia, Where's Russia? A Cartogram of the World's Population

A cartogram is a map on which the actual geography is distorted in order to demonstrate information about the region shown. This cartogram shows information about the world's population, with each country's size weighted to reflect the number of its inhabitants. The discrepancies between your average standard world map and this one are obvious—the obviousness of the distortion being a good indicator of how good a cartogram is.

For example, on a normal world map, Australia (3 million square miles; 7.7 million sq. km) would dwarf Indonesia (741,000 square miles; 1.9 million sq. km). Yet the opposite happens here. Oz might be big, but it's a Big Empty, holding no more than 20.5 million people (2006 est.) Meanwhile, the emerald archipelago to Australia's Near North is teeming with 223 million people (2005 est.), enough to fill eleven Australias. That imbalance is reflected well in this map. Australia almost drowns in the ocean, just like that other sparsely populated "West-

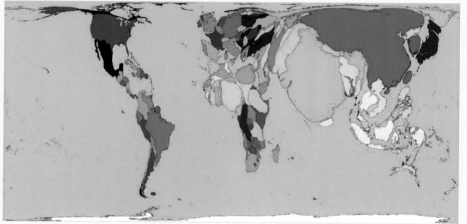

Total Population

In Spring 2000 world populat[ion] estimates reached 6 billion; t[he] thousand million. The distribu[tion] the earth's population is show[n on the] map.

India, China and Japan appea[r large] on the map because they hav[e large] populations. Panama, Namib[ia,] Guinea-Bissau have small pop[ulations,] so are barely visible on the m[ap.]

Population is very weakly rela[ted to] land area. However, Sudan, w[hich is] geographically the largest co[untry in] Africa, has a smaller populati[on than] Nigeria, Egypt, Ethiopia, Dem[ocratic] Republic of Congo, South Afr[ica or] Tanzania.

The size of each territory shows the relative [proportion] of the world's population living there.

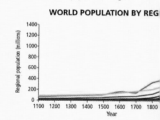

Land area

Technical notes
- Data source: United Nations Development Programme, 2004, Human Development Report.
- Population data is from 2002
- The population net included is estimated as 2 to 3 million (see Appendix map 2).
- See website for further information.

MOST AND FEWEST PEOPLE

Rank	Territory	Value	Rank	Territory	Value
1	China	1295	191	Saint Kitts & Nevis	42
2	India	1050	192	Monaco	34
3	United States	291	193	Liechtenstein	33
4	Indonesia	217	194	San Marino	27
5	Brazil	176	195	Palau	20
6	Pakistan	150	196	Cook Islands	18
7	Russian Federation	144	197	Nauru	13
8	Bangladesh	144	198	Tuvalu	10
9	Japan	128	199	Niue	2
10	Nigeria	121	200	Holy See	1

millions   *thousands*

WORLD POPULATION BY REG[ION]

"Out of every 100 persons added to the population in the coming decade, 97 will live in developing coun[tries"]

ern" outpost in the Far East, New Zealand.

A similar reversal of roles exists between Russia (6.6 million square miles; 17 million sq. km; 142 million inhabitants) and China (3.7 million square miles; 9.6 million sq. km; 1.3 billion inhabitants). The population map reduces Russia to a thin sliver of land, insignificant compared to the giant that is China, which dwarfs just about any country far or close by, except India. Together, these two Asian countries account for fully one-third of the world's population. Incidentally, the number of Indians is slated to surpass China's population later this century.

The map similarly illustrates Canada's relationship with its "bigger" neighbor to the south. Elsewhere, regional dominances become more apparent also. Ethiopia dominates East Africa, Nigeria is by far the largest country of West Africa—in fact, the largest of all of Africa, larger than Sudan, which is huge and empty. The preliminary results of a recent Nigerian census seem to indicate a population of about 140 million people, indeed surpassing by far Africa's second most populous nation, Egypt.

This map also allows for quick "guesstimates" of which countries have an equally large population. The matches can be instructive and surprising. France and Egypt seem about the same size, as are Germany and Ethiopia. Ireland is more or less the same size as Haiti.

One obvious discrepancy is that the map shows Antarctica, which has a permanent indigenous population of exactly zero, and therefore should not be shown.

8

Staring at the Sun: All Total Solar Eclipses Until 2020

Europe is not a good place to witness total solar eclipses, at least not until 2021—unless you live on the Faroe Islands or Spitsbergen, in which case a total solar eclipse will swing by on March 20, 2015. Past consolation prizes since 2000 were total solar eclipses crossing Abkhazia (and other parts of western Georgia) and the Russian Caucasus on March 29, 2006, and the Russian Arctic islands of Novaya Zemlya on August 1, 2008. On May 31, 2003, a less spectacular annular solar eclipse could be observed in the skies over Iceland, parts of Greenland, again the Faeroes, and the northern tip of Britain. Another annular eclipse crossed the Iberian Peninsula on October 3, 2005.

During a total solar eclipse (marked in blue on the map), the moon passes between the earth and the sun, totally obscuring the sun and leaving only a faint corona surrounding the moon. During an annular solar eclipse (in red), the apparent size of the moon is smaller than during a total solar eclipse, leaving a much brighter edge around the moon (an annulus, or ring).

North Americans will get one each: an annular eclipse will cover a band from western Texas, via New Mexico, Arizona and Nevada to Northern California on May 20, 2012, and a total eclipse will culminate near Memphis on August 21, 2017. In South America, the best place to be is Argentina, where total eclipses will pass on July 2, 2019 (its band just touching the Buenos Aires area), and on December 14, 2020 (northern Patagonia). Other eclipse hotspots of the first two decades of this century are Africa and Indonesia.

Solar eclipses occur up to four times a year, and yet total (or annular) eclipses are very rare: They exist only in the relatively narrow strip traced by the moon's shadow (or umbra) on the earth. A partial eclipse can be seen in a much broader area surrounding this strip. The shape of these strips varies according to their position on the map—curving slightly close to the equator, almost circular nearer the poles. Their direction changes according to the seasons.

Total and Annular Solar Eclipse Paths: 2001 — 2020

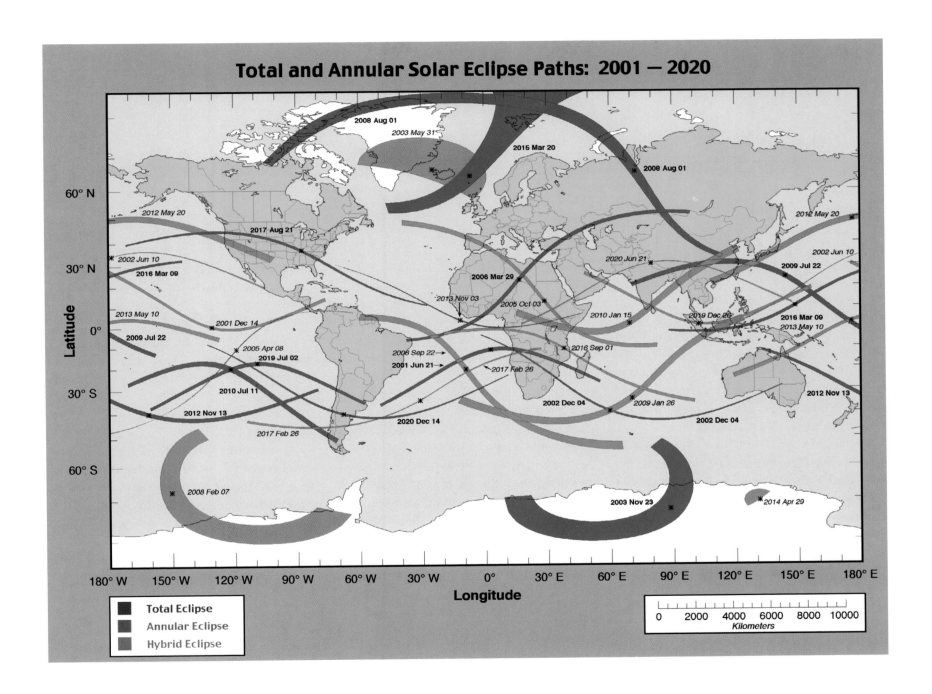

Latitude

60° N

30° N

0°

30° S

60° S

Longitude

180° W 150° W 120° W 90° W 60° W 30° W 0° 30° E 60° E 90° E 120° E 150° E 180° E

2008 Aug 01
2003 May 31
2015 Mar 20
2008 Aug 01
2012 May 20
2012 May 20
2017 Aug 21
2002 Jun 10
2016 Mar 09
2020 Jun 21
2002 Jun 10
2006 Mar 29
2009 Jul 22
2013 May 10
2013 Nov 03
2005 Oct 03
2010 Jan 15
2019 Dec 26
2016 Mar 09
2013 May 10
2001 Dec 14
2009 Jul 22
2005 Apr 08
2019 Jul 02
2006 Sep 22
2016 Sep 01
2001 Jun 21
2017 Feb 26
2010 Jul 11
2002 Dec 04
2009 Jan 26
2012 Nov 13
2012 Nov 13
2020 Dec 14
2002 Dec 04
2017 Feb 26
2008 Feb 07
2003 Nov 23
2014 Apr 29

Legend:
- Total Eclipse
- Annular Eclipse
- Hybrid Eclipse

Kilometers
0 2000 4000 6000 8000 10000

9

Even Penguins Can Go Online: Country Code Top-Level Domains

The two or three letters behind the final dot in any e-mail or Web address are called a top-level domain name (TLD). There are just over 260 TLDs, the most popular of which is .com, one of about a dozen generic TLDs (other ones include .org, .net and .edu). In all, 246 of these extensions are assigned geographically, to each UN-recognized country and to several nonsovereign islands and territories.

These country code top-level domain names (ccTLDs) account for less than half of over 140 million TLDs now in use. Unlike nongeographic TLDs, ccTLDs always consist of two letters only. Most abbreviations are easily recognizable—.br for Brazil, .us for the United States, .in for India—with the exception of those derived from country names in native or other languages that differ from their English variant—.de for Deutschland (Germany), .ch for Confoederatio Helvetica (Switzerland), .hr for Hrvatska (Croatia), and

so on. And then there are those countries or territories just too small to be well known— .fo for Faroe Islands (an autonomous Danish archipelago north of Scotland), .bv for Bouvet (an uninhabited Norwegian island closer to Cape Town than to Oslo), .hm for Heard and McDonald Islands (an uninhabited Australian possession in the southern Indian Ocean).

This map presents all the ccTLDs in their geographic context, and in a size relative to the population of each country or territory —except for China and India, which were restrained by 30 percent to fit the layout, and for all those uninhabited islands, which wouldn't have appeared on the map other-wise. The smallest type reflects countries with 10 million inhabitants or fewer. The ccTLDs are color-coded by region (Americas, Europe, Africa, Middle East, Asia-Pacific).

Some of the more intriguing abbrevia-tions are:

.ax (Aland)
.aq (Antarctica; glad somebody remembered the penguins)
.im (Isle of Man)
.io (British Indian Ocean Territory)
.sm (San Marino)
.tf (French Southern and Antarctic Lands)
.tl (East Timor)
.pm (Saint-Pierre and Miquelon)

Some small countries have made a tidy profit because their country code can be used as a vanity ccTLD, often leading to strange virtual associations. Some examples:

.ad (Andorra): advertising agencies.
.ag (Antigua and Barbuda): agricultural sites. In Germany, AG (short for *Aktiengesellschaft*) is appended to the name of a stock-based company, similar to Inc. in the United States.

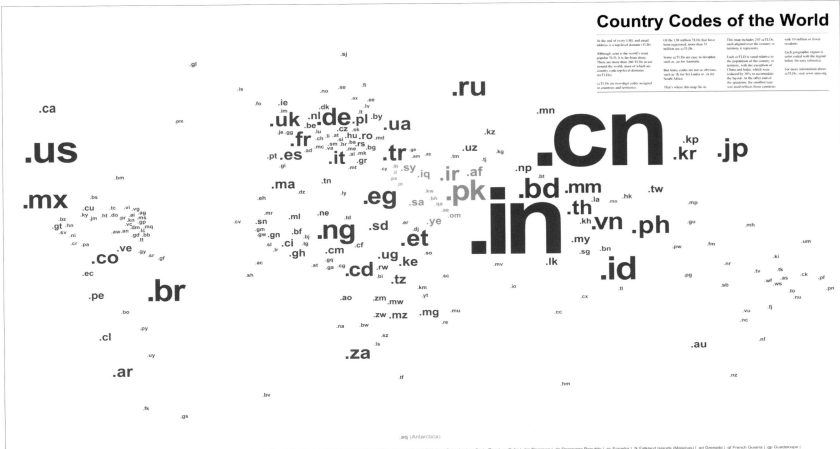

.am (Armenia): AM radio stations.

.cd (Democratic Republic of the Congo): CD merchants and file-sharing sites.

.dj (Djibouti): disk jockeys.

.fm is a ccTLD for the Federated States of Micronesia but it is often used for FM radio stations.

.gg (Guernsey): gaming and gambling industry, particularly in relation to horse racing ("gee-gee").

.la (Laos): marketed as the TLD for Los Angeles.

.tv is a ccTLD for Tuvalu but it is used by the TV/entertainment industry.

.vu is a ccTLD for Vanuatu but means "seen" in French.

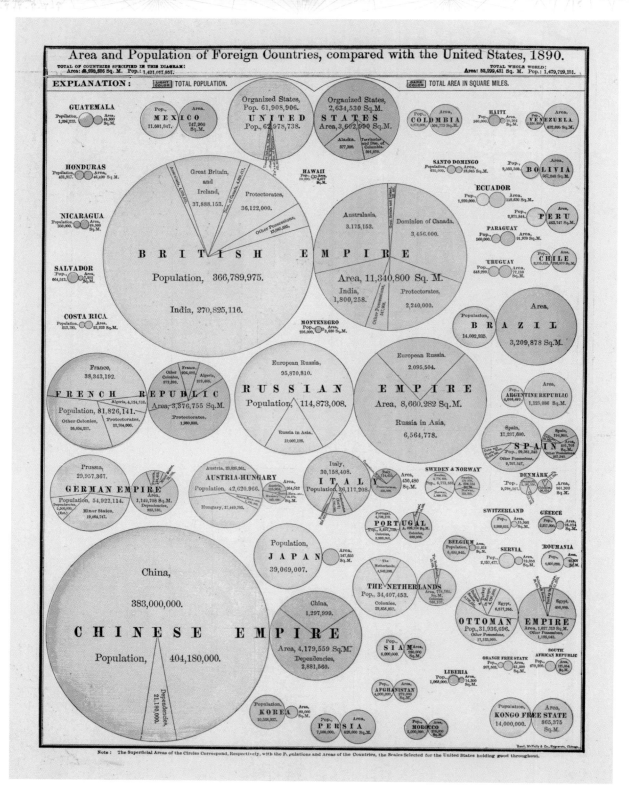

Area and Population of Foreign Countries, compared with the United States, 1890.

10

Bubbles Bursting: Area and Population of Foreign Countries (1890)

The size of most countries hasn't changed much since 1890, but boy did we have a bit of a growth spurt populationwise. In 1890, there were just under 1.5 billion people on earth. That number has more than quadrupled to today's 6.6 billion-plus. The seven billionth human is expected in 2012.

What's changed is more than just numbers: Europe amounted for about 25 percent of the world's population back in 1890; today it represents no more than 10 percent and has been surpassed by Africa (around 15 percent; less than 10 percent in 1890). Asia, for centuries the most populous continent, has gotten even more important, climbing from around 57 percent to just over 60 percent. Latin America, at 5 percent around 1890, is edging toward 10 percent. North America,

also starting out at 5 percent, has stabilized at that percentage.

Then as now, China was the world's most populous nation, but its population of just over 400 million would more than triple to 1.3 billion today. The second most populous nation was the British Empire (366 million), in large part thanks to India (270 million, now 1.1 billion). The number of inhabitants of Great Britain and Ireland then was 38 million, their combined populations now standing at 65 million.

That's not even double the population of over a century ago—compare that with the tenfold increase in the number of Mexicans (11 to 110 million) or the *thirteenfold* multiplication of Brazilians (14 to 184 million). This almost obscures the impressive—and

numerically more than twice as important—growth in the United States, which saw the number of its inhabitants increase from 62.5 to 304 million.

An updated version of this remarkable population map would still be dominated by China, with India—having broken free from the British Empire—a close second. The midsized representations of European nations would be much smaller—especially that of the Netherlands: This 1890 map still includes the over 34 million inhabitants of its colonies, mainly Dutch East India (now Indonesia, and at 235 million inhabitants now the fourth most populous nation, after China, India and the United States). The Netherlands proper only barely managed to triple its population (from 4.5 million to a still puny 16 million).

11

The Inglehart-Welzel Cultural Map of the World

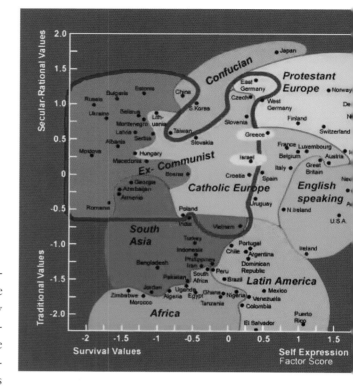

East and West Germany are next to each other, as one would expect. But Romania's closest neighbor is Armenia? And Poland and India are side by side? Well, this is not a straightforward geographical map, but a cultural one. It plots out how countries relate to each other on a double axis of values (ranging from "traditional" to "secular-rational" on the vertical and from "survival" to "self-expression" on the horizontal scale). This makes for some strange bedfellows: For example, South Africa, Peru and the Philippines occupy almost the same position, although they're on three different continents.

Ronald Inglehart, after whom this map is half-named, is a political scientist at the University of Michigan and director of the World Values Survey, which charts cultural differences and changes all over the world. The two dimensions mentioned earlier ("traditional/ secular-rational"and"survival/self-expression") apparently explain more than 70 percent of cross-national variance in ten indicators.

Four survey-waves have been executed be-tween 1981 and 2001 in eighty societies. In-glehart's work demonstrates significant value shifts—and predictable ones at that—especially in those societies moving through a late indus-trial or to a post-industrial phase. One of those changes is the diminishing role of gender differ-ences, but the predictability extends to attitudes toward religion, politics and family life.

For example, in societies near the "tradi-tional" side of the traditional/secular-rational axis, religion is very important. This usually always implies a strong emphasis on family values, deference to authority, rejection of abortion, divorce, euthanasia and suicide, and even seems to predict a very nationalistic out-look on life. In countries more to the "secular-rational" side of this axis, the attitudes toward these topics is reversed.

The other axis represents the shift from a society dominated by the struggle for survival to one where survival is a given and the em-phasis of the "struggle" is on subjective well-being, quality of life and self-expression.

These shifts from a materialist toward a postmaterialist culture should eventually lead to less dirigiste, more democratic societies. And to less religious ones too, consistent with the thesis that an increase in secularism is a by-product of this development. This might have seemed to be the trend throughout most of the twentieth century, but that trend has arguably reversed in recent years, in the Mus-lim world as in the Americas, among others (Europe still being a notable exception). In-glehart points out that secularism coincides with dramatically falling birthrates, thus explaining why the "triumph" of secularism seems to be accompanied by a rising tide of religious traditionalism and fundamentalism: People in those categories constitute a grow-ing proportion of the world's population.

XVII. MAPS FROM OUTER SPACE

Boldly going where no atlas has gone before.

1

A Foldable Map of Mars's Moon Phobos

The planet Mars has two moons, Phobos ("fear") and Deimos ("dread"), named after the God of War's horses, sons or servants (opinions vary). Phobos is the larger and closer of these two, but is still so small as to be porous rather than solid, and to have a highly irregular shape.

To see exactly how irregular Phobos's shape is—more potato than orange—you need only this map, a pair of scissors and some glue or tape. Chuck Clark, an architect, has been experimenting with something he calls "constant-scale natural boundary mapping," a method for making a two-dimensional map of an object, even a highly irregular object like Phobos, which may be folded into a realistic facsimile of the three-dimensional object.

Phobos is 17 × 13 × 11 miles, with a surface of 3,790 square miles. Contributing to the moon's irregular shape is Stickney Crater,

the result of a giant impact that nearly broke up Phobos.

Phobos will probably break up in the future: Its orbit is lowering at the rate of 59 feet per century, meaning that Martian gravity will pulverize it into a planetary ring in 30 to 80 million years' time.

Before spacecraft sent images back, showing Phobos to be a natural object, its strange shape and orbit had some scientists speculate that it might be a hollow object, possibly even an artificial base circling Mars.

Phobos has been suggested as a landing site for a future manned Mars mission; its low gravity would make landing and taking off considerably easier than on Mars itself. The Russians plan to launch Phobos-Grunt in 2009, a mission that would include a landing on the moon and the return of soil samples (*grunt* is Russian for "soil") to earth.

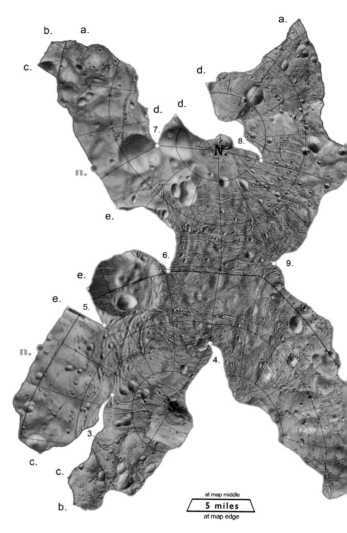

at map middle
5 miles
at map edge

2

The Colorful Side of the Moon

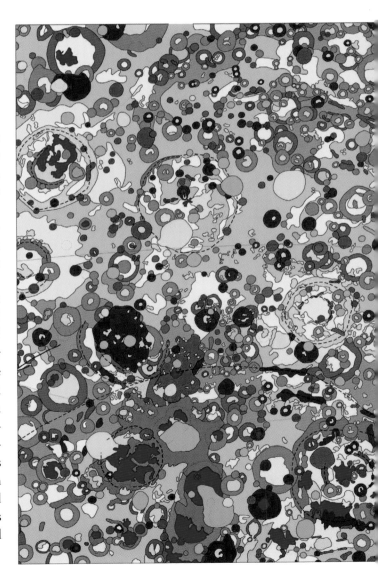

The moon is tidally locked to the earth, so it always faces us with the same side—although a process called libration ("swaying") makes it such that over time we can actually see about 59 percent of our natural satellite's surface.

The other side of the moon should be called the far side, *not* the dark side. Both sides of the moon receive sunlight equally; it's only during a full moon (when the half facing the earth is illuminated) that the far side of the moon corresponds to its dark side.

I blame Pink Floyd's *The Dark Side of the Moon* (1973) for popularizing the erroneous term, although I concede that as concept album titles go, *The Central Far Side of the Moon* sounds slightly less grand. And yet how riotously colorful doesn't this map of that selfsame central far side of the moon look?

NASA and the U.S. Geological Survey have produced a series of maps of the geological structure of several heavenly bodies: the moon, Mercury, Mars, Venus and a few others in our solar system of which probes have sent back geological information. The study of the geology of other heavenly bodies is called exogeology (or astrogeology) and, in the specific case of the moon, sometimes selenology.

Selenology has its points of comparison with action painting, for these energetic splashes look like something out of the Jackson Pollock School of Hurling Paint at a Canvas.

In fact, a lot of hurtling was involved in the creation of this map. The circularity of many blotches reflects the large number of meteorite impacts on the lunar surface over many, many centuries. Unlike earth, the moon doesn't have an atmosphere that can burn up all but the bigger rocks crashing in on its surface. The different colors denote different types of soil. Pink stands for the material of grooves and mounds, covering craters and other features of pre-Nectarian through Imbrian age. Green stands for the material of sharp-rimmed craters, while yellow stands for the material of *very* sharp-rimmed rayed craters. And so on.

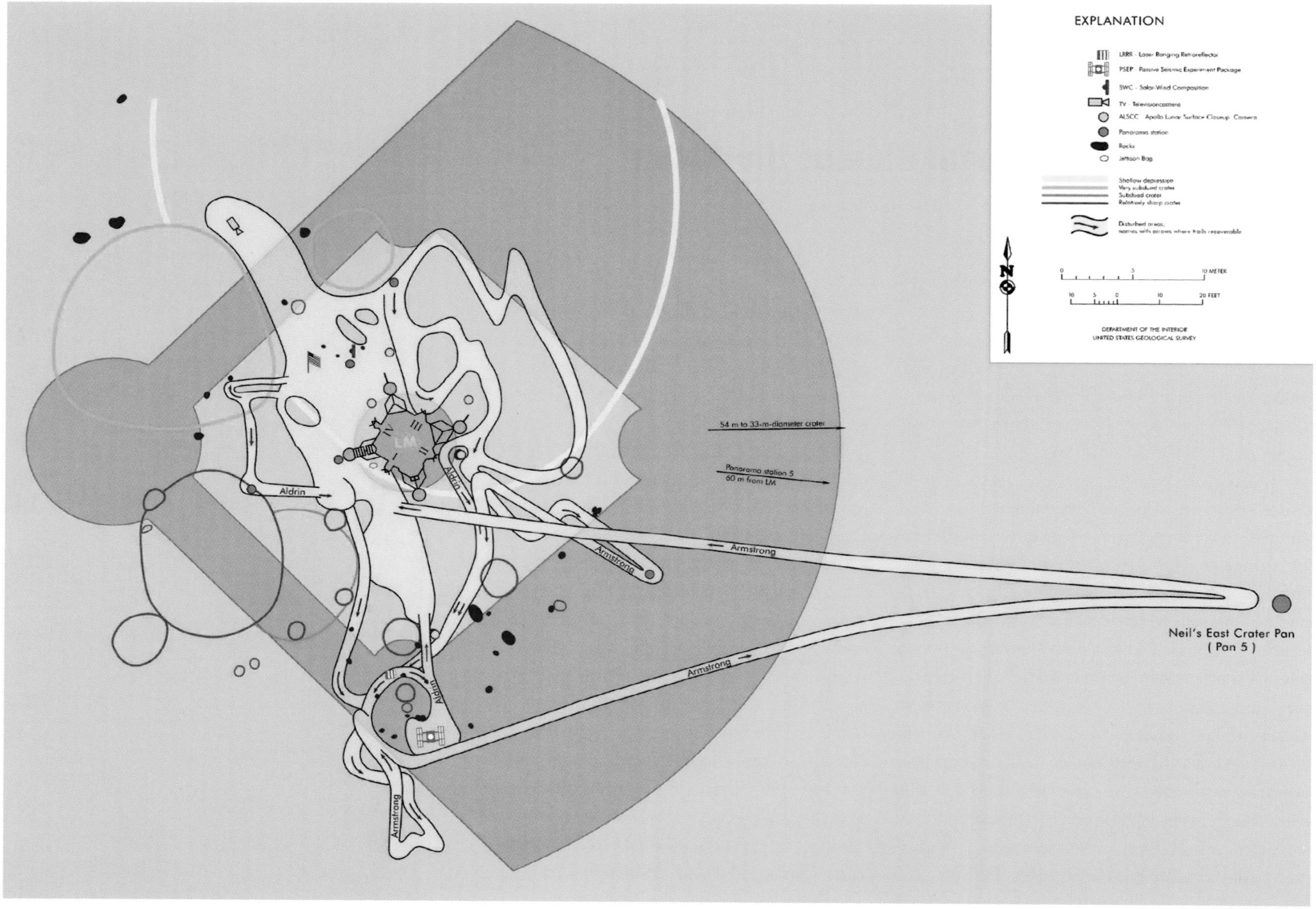

EXPLANATION
LRRR - Laser Ranging Retroreflector
PSEP - Passive Seismic Experiment Package
SWC - Solar-Wind Composition
TV - Televisioncamera
ALSCC Apollo Lunar Surface Closeup Camera
Panorama station
Rocks
Jettison Bag
Shallow depression
Very subdued crater
Subdued crater
Relatively sharp crater
Disturbed areas;
names with arrows where trails recoverable
N
0 5 10 METER
10 5 0 10 20 FEET
DEPARTMENT OF THE INTERIOR
UNITED STATES GEOLOGICAL SURVEY
54 m to 33-m-diameter crater
Panorama station 5
60 m from LM
LM
Aldrin
Aldrin
Aldrin
Armstrong
Armstrong
Armstrong
Armstrong
Neil's East Crater Pan
(Pan 5)

3

One Small Stroll for Man: The First Moonwalk

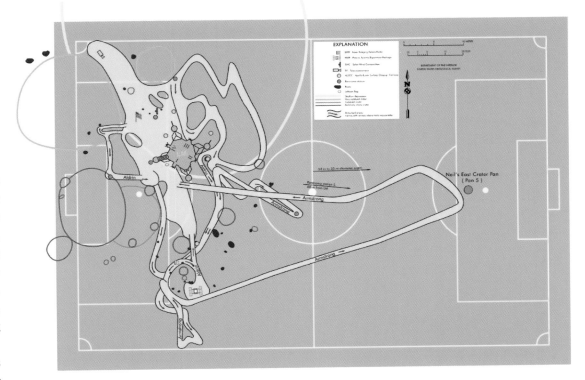

On July 21, 1969, Neil Armstrong stepped out of the *Eagle*, spoke the historic words, "That's one small step for [a] man, a giant leap for mankind," and became the first human to walk on the moon. He didn't moonwalk alone—"Buzz" Aldrin joined him on the surface—and he didn't walk far.

In fact, the giant leap for mankind looks more like one small stroll for man on these maps. After traveling hundreds of thousands of miles, the landing crew of the Apollo 11 lunar mission spent two and a half hours on the lunar surface and in that time barely covered an area the size of a baseball diamond. Or, if your sports preferences lie elsewhere, half a soccer field, with Armstrong making just one dash at the other side's goal.

Obviously, the object of the mission was to prove only that a manned moon landing could be made, and the crew safely returned to earth. Subsequent missions did stray much farther afield—the later Apollo missions even had Lunar Rovers (dune buggies for astronauts), allowing for a much wider range.

Juxtaposing these two maps shows up a curious discrepancy between the paths traced on each. Fodder for those who believe the landings never happened? "The two renditions of Neil's little trip out to East Crater (also known as Little West Crater) differ somewhat, primarily because we don't know exactly what route Neil followed," says Eric Jones, a historian involved in creating these maps. "We should have better info when NASA's Lunar Reconnaissance Orbiter reaches the Moon and starts taking data later in 2008."

4

Naming Titan's Methane Sea

Except for some of the harsh, imperma-nently inhabited and sparsely visited in-lands of Kerguelen, there are no places left on earth to name. Those with a penchant for baptizing should look to the priesthood, or take a more literal interest in heaven—there are ever more known worlds out there, and precious little of those exoplanets have been explored, let alone provided with toponyms. Even within our own solar system, the field is still wide open. Although all planets and major moons in our solar system have been named, many of their geograpical features haven't.

Titan, Saturn's largest moon, was discov-ered in 1655 by Christiaan Huygens. Titan is larger in diameter than the smallest planet, Mercury, and 50 percent larger than our own moon. It is the only moon in our solar system to have a dense atmosphere—so dense that, in combination with its limited gravity, humans on Titan could fly by just flapping their arms.

The orange opacity of Titan's atmosphere makes the moon appear bigger than it actu-

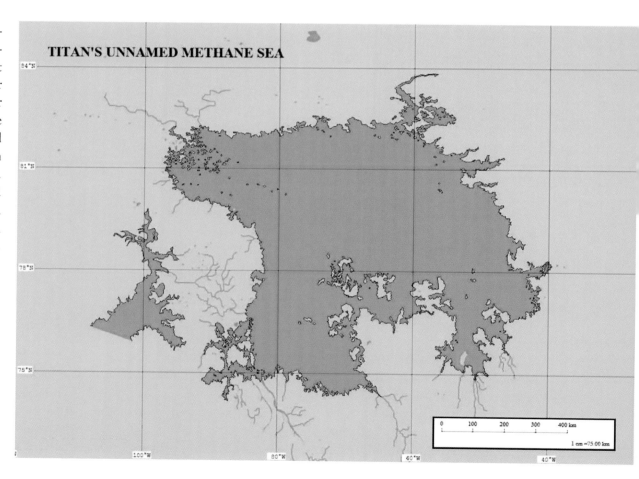

TITAN'S UNNAMED METHANE SEA

ally is—astronomers have since distinguished between permanent cloud cover and surface, and downgraded it from the first to the second largest moon in our system, after Jupiter's satellite Ganymede.

Not until the fly-by, in 2004, of the Cassini-Huygens mission could scientists confirm the speculation, first ignited by both Voyager missions and then heightened by Hubble observations, that Titan is the only heavenly body (save earth) to contain large liquid surfaces—seas or lakes, as non-astronomers would call them. For they seem a bit too small to be labeled oceans. These seas most probably consist of methane or another hydrocarbon.

Cue Peter Minton, a teacher from San Diego with a thing for maps, for mapping and more precisely still for mapping shorelines. His Web site shows the many maps he has made, most with more conventional subjects than these seas on Titan. Mr. Minton used the data at the Jet Propulsion Laboratory to create this map of Titan's Unnamed Methane Sea, detected in mid-2006 by the Cassini probe.

Fascinating. It looks a bit like the Aral Sea, although that might just be me confusing this color scheme with satellite pictures of the Central Asian lake, shrinking into the desert. The many rivulets and islets make it look like a nice lake to vacation at, until you remember that there's something unpleasant in the air there. A shame: How nice it must be to flap your arms and fly over this Superior-sized lake. But then again, the sunlight hardly penetrates Titan's cloud cover, so you wouldn't see much. And the average temperature is −180°C (−290°F). Can we go home now?

The naming bit, then. The unnamed methane lake *has* recently been named by the International Astronomical Union. It's now officially called Ligeia Mare, after a siren of Greek mythology ("Ligeia" is also the title of a gothic short story by Edgar Allan Poe). The USGS *Gazetteer of Planetary Nomenclature* states that large methane seas on Titan will be named after mythological or literary sea creatures.

Other suggestions for the lake's name were received before Ligeia was adopted by the IAU:

- "It looks like Amy Winehouse's hairstyle. How about Aquanet Lacus?"
- "We could call it Lake Kerguelen."
- "How about Lake Cocytus, after the deepest (and frozen-over) circle of Hell in Dante's Inferno? If that isn't already taken."
- "I thought that big one was going to be called Lake Vonnegut."
- "Lake Gore."
- "It looks like a turtle jumping a hurdle to me."

XVIII. WATCHAMACALLIT

Some maps defy categorization—at least in the context of the other, subjective categories in this book.

1

True or False: The Vinland Map

The Vinland Map was discovered in 1957, bound up with a manuscript of undisputed antiquity, the *Historia Tartorum*. The map supposedly is a fifteenth-century copy of a thirteenth-century world map, showing the known parts of Europe, Asia and Africa, as well as an unknown land across the Atlantic Ocean labeled Vinland, which it claims was visited in the eleventh century. The map mentions that "by God's will, after a long voyage from the island of Greenland to the south toward the most distant remaining parts of the western ocean sea, sailing southward amidst the ice, the companions Bjarni and Leif Eriksson discovered a new land, extremely fertile and even having vines, . . . which island they named Vinland."

This corresponds well with a tradition in Viking folklore that Norsemen, using Iceland and Greenland as stepping stones, had a more or less regular contact with North America. According to some Icelandic sagas, North America was sighted in about 986 by Bjarni Herjolfsson, who was blown off course on

a trip from Iceland to Greenland. His stories lured Leif Eriksson on an expedition in the year 1000, on which he named (north to south):

- Helluland ("Flatstone Land"—possibly present-day Baffin Island);
- Markland ("Wood Land"—possibly Labrador); and
- Vinland ("Grapevine Land" or "Pasture Land"—possibly Newfoundland).

Eriksson established two settlements: Straumfjördr in the north and Hop in the south. Both attempts failed quickly, also due to attacks from the native *skraelingar* ("barbarians"), and were never repeated.

In 1960, archaeological excavations at L'Anse-aux-Meadows on Newfoundland turned up the remains of a Viking camp. For the first time, scientists established that Vikings actually did cross the Atlantic. Interest in all things Vinland soared. Yale University bought the map in 1965, had it insured for

$25 million and published it that same year. That was the starting point for two debates that rage to this day: Where is Vinland? And: is the map real?

Some have placed Vinland in New England; after 1960 many were sure it was at L'Anse-aux-Meadows, thinking any location more to the south unlikely. Others postulated that L'Anse was an undocumented colonization attempt, leaving open the possibility of Vinland having been more to the south, some would say as far south as present-day Rhode Island.

The map's authenticity was maintained by Yale at its second edition in 1995. Which is remarkable, considering the amount of criticism it has had to endure. Such as:

- While the map has been radiocarbon-dated to between 1423 and 1445, it appears to have been coated with an unknown substance in the 1950s. This could be an undocumented attempt at preservation, or it could be part of

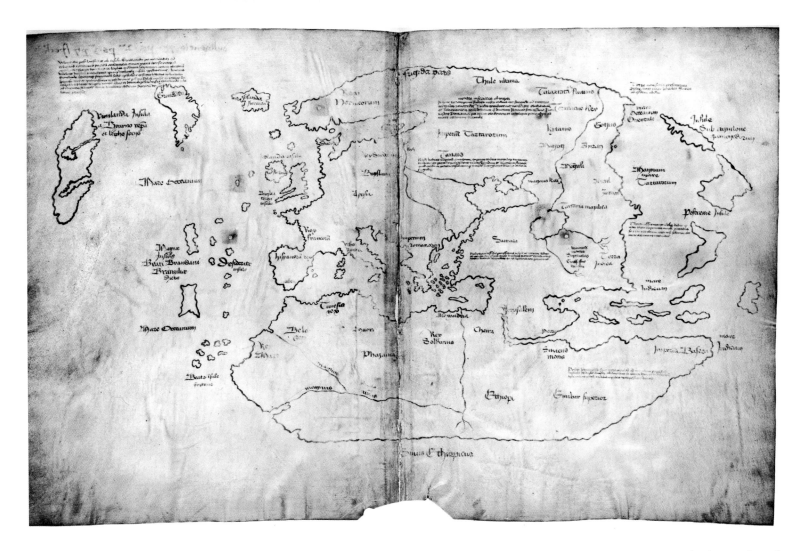

a forger's attempt to draw a new map over an old one. It's unclear whether this substance is over or under some of the ink on the page.

- The ink itself has been chemically analyzed, and dated to after 1923 due to the presence of anatase—a synthetic pigment in use only since the 1920s.

Natural anatase has been demonstrated in various medieval manuscripts, though.

- As for the content of the map, a number of questions challenge the age of the document. Greenland is presented as an island—a fact not physically proven until the turn of the twentieth century and unknown to the Vikings, who mostly

thought it a peninsula descending from the north.

- Several passages in the text are equally anomalous.
- Finally, the best argument against the map's veracity seems to be that the Vikings were such good seafarers that they didn't use nautical maps at all.

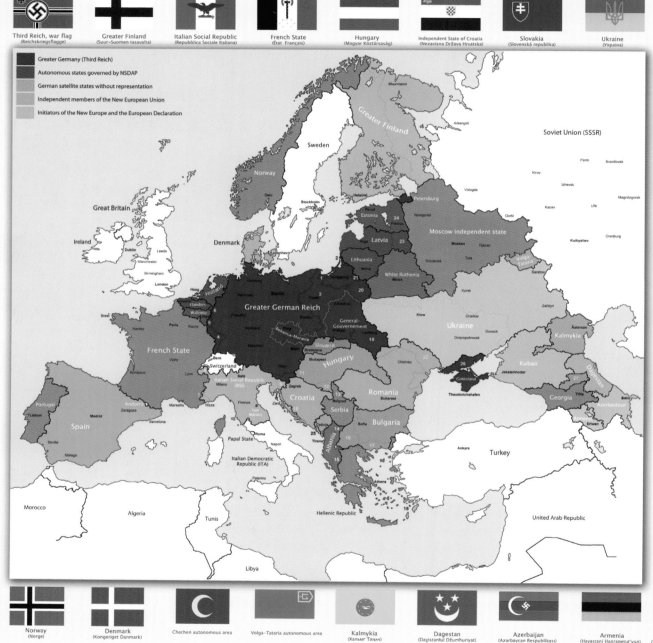

2

Europe, if the Nazis Had Won: Neuropa

What if the Nazis had won the Second World War? For some time after the German blitzkrieg victories over Poland (September 1939) and in western Europe (May–June 1940), the question seemed anything but hypothetical.

And yet the Nazis themselves don't seem to have had a clear plan for organizing Europe or the world after their victory. No contemporary maps survive of what a postwar Nazi Europe would look like. Or they have been overlooked, which would seem strange for the most intensely studied conflict in world history.

That no such maps exist, might be because the Nazis were less organized than the stereotype of the "obedient, efficient German" implies. Nazi rule was a chaotic mess of competing power centers, all vying for the Führer's approval.

Which leaves speculation about a postwar Nazi Europe (or world) up to alternate (or speculative) history, either a maligned branch of history or an obscure branch of

science fiction. Some examples of the latter include Robert Harris's novel *Fatherland*, set in 1960s Nazi Europe, and Philip K. Dick's *The Man in the High Castle*, portraying life on the Japanese-ruled U.S. West Coast.

This map does give what seems to be a well-considered vision of a Europe-wide Nazi state in the 1980s, as it might have emerged after a German victory. German supremacy is "concealed" by the construct of Neuropa ("New European Union"), a sort of evil twin of the European Union in this universe. Unclear is which is the point of divergence (POD), when "real" events split off into the "alternate" version of history leading to this map. Possible PODs might be a German victory over Russia and/or a failed Allied invasion of Normandy. An overview:

A. Greater Germany
This Grossdeutsches Reich

This largely corresponds to the Third Reich as it existed at the height of its power:

- Germany anno 1937, enlarged by the *Anschluss* of Austria (*Ostmark* in Nazi parlance) and the annexation of the Sudetenland area of Czechoslovakia, both in 1938; and the "protectorate" of Bohemia-Moravia, which Germany occupied in early 1939. It also includes the Memel area (2), Hitler's last territorial acquisition (from Lithuania) before the start of the war;

- Areas Germany conquered in Poland: the Free City of Danzig (1) and Wartheland (3), following its September 1939 invasion of Poland. Also included are Sudauen (21), a district conquered by the Soviets but transferred to Germany as part of the Molotov-Ribbentrop Pact, and the General Gouvernement (again as a so-called protectorate), Galicia (19) and Bialystok (20), Polish areas conquered from the Soviets at the beginning of Operation Barbarossa.

- Areas conquered during the blitzkrieg in the west: Alsace-Lorraine (4) from

France, the formerly independent Grand Duchy of Luxembourg (5) and the Eupen-Malmédy area (6) of Belgium.

- The sum of all the preceding territories adds up to the Third Reich as it really existed at the height of its power. Counterfactual additions on this map are Petersburg, the former Russian capital on the Gulf of Finland, and Gotenland (26), a German colony encompassing the Crimea and other parts of southern Ukraine.

B. Autonomous States Governed by the NSDAP

Here, the map's color scheme gets a bit confusing. Allowing for slight differences within each color, these states, governed by Hitler's National Socialist Party, should be:

- In the west: the Netherlands, Flanders and Wallonia. Both successor states to Belgium acquire a chunk of northern France. All three are designated as a *Reichsgau*.
- In the east: Lithuania; Latvia, enlarged with Nevel (23), the former Russian oblast of Pskov; Estonia, enlarged with formerly Russian territories of Pleskau and West Ingria (24); and White Ruthenia, almost identical to present-day Belarus. All four areas are designated as a *Generalbezirk*.

As these countries all immediately border Greater Germany and contain large German (or at least Germanic) populations, it's possible that NSDAP rule is the preamble for an eventual complete absorption into the Reich.

C. German Satellite States Without Representation

These would seem to be concentrated only in the east and include (again allowing for variation within this color scheme): a "Moscow Independent State" and Volga-Tataria in northern Russia, and Kalmykia, Kuban, Dagestan, Georgia, Armenia, Azerbaijan and Chechnya-Ingushetia (27) in southern Russia and the Caucasus.

D. Independent Members of the New European Union

- In the west: Spain and Portugal, which already before the war had Nazi-friendly regimes (and aren't subject to any border changes); and France, losing bits to Flanders, Wallonia, Germany and Italy.
- In the north: Norway and Denmark, with the Norwegians ceding some land in the extreme north to the Finns.
- In central Europe and the Balkans: Slovakia; Croatia, including much of Bosnia; Hungary, having annexed part of present-day Romania and Hungarian parts of Slovenia (11) and Croatia (12);

Serbia, including the autonomous Banat region (13); Montenegro (25); Albania, enlarged with parts of Serbia's Kosovo (14) and Greece's Epirus and Pindus region (15); Greece, minus that and other regions; Bulgaria, plus Macedonia (16) and western Thrace (17); and Romania, receiving Transnistria (22) in exchange for the loss of Transylvania to Hungary.

E. Initiators of the New Europe and the European Declaration

- Greater Finland: including a bit of Norway and a lot of Russia, basically everything between Lake Ladoga and the northern port of Murmansk.
- The Italian Social Republic: covering northern Italy, enlarged with Nice (7) and Savoy (8) from France, part of Slovenia (9) and Dalmatia (10) from Yugoslavia and the Dodecanese Islands (18) from Greece.

The southern half of Italy is made up of an Italian Democratic Republic (and a papal state), which by the look of it are neutral, as are Switzerland, Sweden, the United Kingdom and Ireland. The status of the surrounding areas in Africa and the Middle East is unclear. Would they all have been neutral? Surely, what remains of the Soviet Union—also in white—wouldn't be. All of which makes this map of a Nazi Europe problematic—fortunately . . .

3

Not Even Faux, Just Plain Wrong: Schoolcraft's Federation Islands

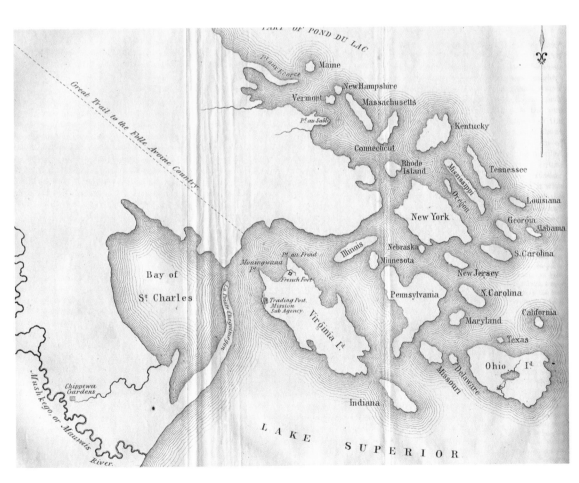

Explorer and ethnologist Henry Rowe Schoolcraft (1793–1864) dotted the Midwest with faux Indian names, conjuring them up from Native, Latin and Arabic syllables. He named the source of the Mississippi Lake Itasca, a portmanteau of *veritas caput*, Latin for "true head." (Schoolcraft had erroneously identified nearby Lake Cass as the river's source on an earlier expedition in 1820).

Schoolcraft was full of not-so-bright ideas. On that same 1820 expedition, he was struck by the fact that an archipelago in Lake Superior, just off the coast of what is now Michigan's Upper Peninsula, seemed to contain the same number of islands as there were U.S. states and territories. He named them after twenty-nine existing states and territories, and collectively dubbed them the Federation Islands.

This turned out to be a real stinker of an idea, not only because the number of states was rapidly increasing at the time (fifteen to twenty states in 1818, twenty-one states in 1819, twenty-three in 1820, twenty-four in 1822), but also because the islands are constantly merging, submerging and remerging due to erosion and the shifting of sandbars—and are thus also constantly changing in number. These two sets of variables made this particular idea not as much faux as just plain wrong.

At some point, Schoolcraft's suggestion was rightly disregarded or mercifully forgotten; the archipelago is now called the Apostle Islands, and all of them have names like Eagle, Sand and Raspberry Island, befitting the protected area they have become. Out of twenty-two Apostles only Madeline Island

is not a part of the Apostle Islands National Lakeshore. Come to think of it, apostles might also be a bit of a misnomer for a collection of anything other than twelve.

This 1820 Schoolcraft sketch of the Federation Islands, oriented toward the west (the north is to the right), shows the twenty-nine islands he named after states and territories. The larger islands are still easily identifiable today: Virginia is now Madeline Island; Ohio, Outer Island; Pennsylvania, Stockton Island; and New York, Oak Island.

Vermont Island has been identified with the now submerged Steamboat Island, and

others too have disappeared beneath the waves or merged with other islands. Three anomalies rack my brain: How would you call an island named after the state of Rhode Island? Rhode Island Island? Why was Michigan not included in Schoolcraft's original list of island names? And why was tiny Indiana island chosen to be renamed Michigan, after the islands' home state?

After his exploring days, Schoolcraft married into the Ojibwe culture and became an expert on its language and lore, providing Henry Longfellow with the material for his epic poem *The Song of Hiawatha*.

4

Clouds with Silver Linings: Europe Discovers the World

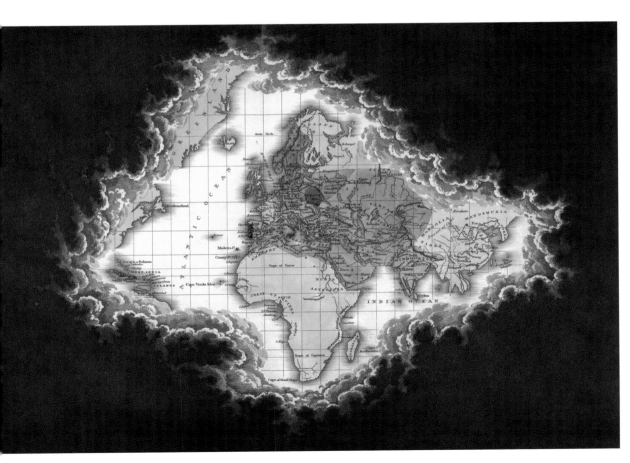

Your standard world map will have the American continent to the left and Asia to the right. Africa's central position is incidental, because this layout is Europe's legacy. No other set of maps illustrates the origin of cartographic eurocentrism better than these two, portraying a world covered by dark clouds, regressing as the European discoveries from the late fifteenth century onward reveal the true extent of the world to the east and west. Europe, the Old World, naturally remains central as new lands are added to the map.

These maps also illustrate the speed with which the Age of Discovery propelled competing European nations to the far-flung corners of the earth. The first map is entitled The Discovery of America and is dated AD 1498, the year of Columbus's third voyage to the "Indies." At that time, the Atlantic coastline of North America is known, some islands in the Caribbean (notably Cuba and Hispaniola) are already under Spanish dominion and the first sketchy reports of the nearby northern coast of South America are in.

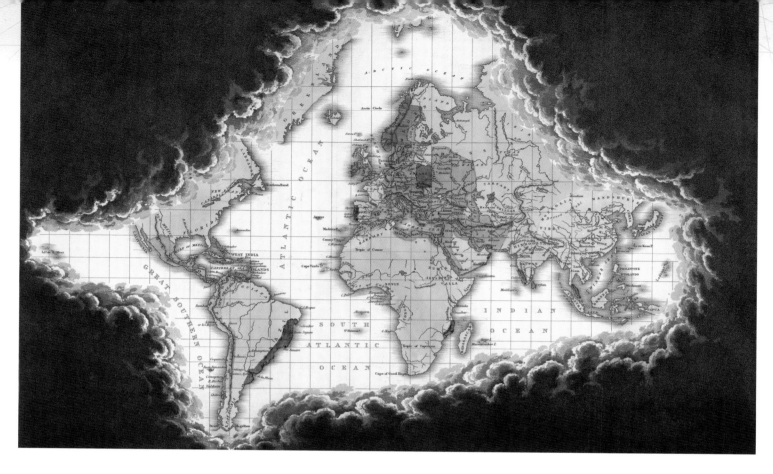

Africa's interior would remain terra incognita for many centuries to come, but even her farthest shores were now known to Europeans. Portuguese seafarers had rounded the Cape of Good Hope in 1488 and found a trade route to India. This had become necessary because Muslim expansion (the green area covering North Africa, the Middle East and Central Asia) had cut off the older overland trade routes between Europe and the East. Europe at this time did not have direct contact with the Far East, but knew about it through an elaborate, trade-fueled version of "Chinese whispers."

The second map is revealing in two respects. It shows the relatively small "extra" area explored by Europeans in Asia at the death of the emperor Charles V in 1558—basically the main islands now constituting Indonesia (some of which were known as the Spice Islands, indicating a main trade motive), and the huge new terra cognita in the Americas: Basically, all but the frozen north and the distant northwest had been explored by European nations, some of which had started colonization of the continent—New France, the nucleus of what could have been a French *Amérique du nord* (and whose legacy lives on in Québec); New Spain, stretching from Mexico down the Andes to Chiloë Island; leaving the east coast of America's south open for the Portuguese projects in Brazil.

Portugal is also present along the searoad east, with colonies in Africa (later to become Mozambique), the Middle East (in Oman and Persia), the Indian Subcontinent (the west coast and Ceylon) and the Malaccan peninsula. The absence of Spanish possessions in the East is no coincidence: The papally endorsed Treaty of Tordesillas (1506) effectively divided the world in two halves, one for Portugal and one for Spain. The dividing line sliced through present-day Brazil, allowing Portugal to maintain its interests in the Americas and monopolize trade with the East.

The Treaty of Zaragoza (1529) later established the antimeridian at the other side of the world, cutting through Japan, New Guinea and Australia, which would remain in the dark for a little while more.

5

Ludacris and the "Ho"-Belt: A Rap Map of "Area Codes"

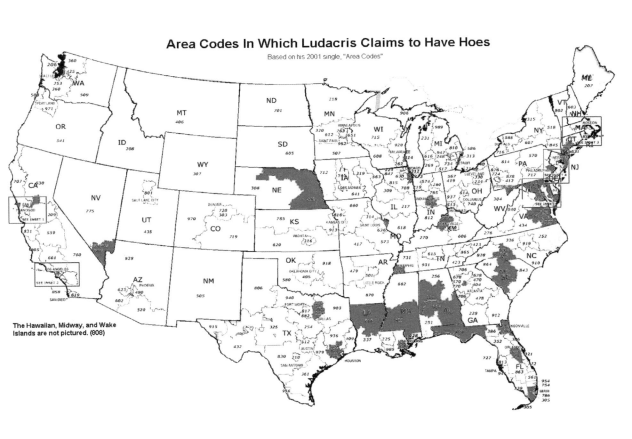

Area Codes In Which Ludacris Claims to Have Hoes
Based on his 2001 single, "Area Codes"

The Hawaiian, Midway, and Wake Islands are not pictured. (808)

'm a female and a feminist. I dislike the usage of the word 'ho.' However, as a geography major, I find this song hilarious, and had to map it," says Stefanie Gray, referring to "Area Codes" by the rap artist Ludacris.

Rap, for those less familiar with the term, is a genre in which the rhythmic delivery of rhyme and wordplay constitutes the main element of the music. Rap relates to singing as racewalking relates to running—but that's just my inexpert opinion.

Rap music has been criticized for its content, which often consists of crude and ludicrous bragging about the rapper's lyrical, financial, criminal, physical and sexual prowess. "Area Codes" could be considered as an example of this phenomenon:

"I'll jump off the G4, we can meet outside
So control your hormones and keep your
 drawers on
'Til I close the door and I'm jumping your
 bones

3-1-2's, 3-1-3's (oh), 2-1-5's, 8-0-three's (oh)
Read your horoscope and eat some
    horderves [*sic*]
Ten on pump one, these hoes is self serve
7-5-7, 4-1-0's, my cell phone just overloads."

"In this song, Ludacris brags about the area codes where he knows women, whom he refers to as 'hoes,' " says Ms. Gray, who plotted out all the area codes mentioned in this song on a map of the United States. She arrived at some interesting conclusions as to the locations of this rapper's preferred female companionship:

- "Ludacris heavily favors the East Coast to the West, save for Seattle, San Francisco, Sacramento, and Las Vegas."
- "Ludacris travels frequently along the Boswash corridor."
- "There is a 'ho belt' phenomenon nearly synonymous with the 'Bible Belt.' "
- "Ludacris has hoes in the entire state of Maryland."
- "Ludacris has a disproportionate ho-zone in rural Nebraska. He might favor white women as much as he does black women, or perhaps, girls who farm."
- "Ludacris's ideal 'ho-highway' would be I-95."
- "Ludacris has hoes in the Midway and Wake Islands. Only scientists are allowed to inhabit the Midway Islands, and only military personnel may inhabit the Wake Islands. Draw your own conclusion."

6

The Antisaloon League Map of the World

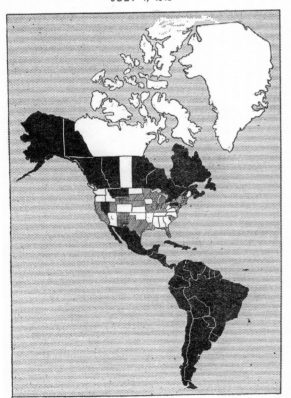

WESTERN HEMISPHERE
WET AND DRY MAP OF WESTERN HEMISPHERE,
JULY 1, 1915

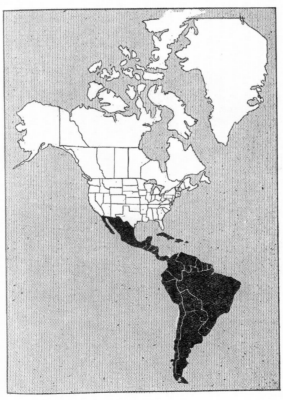

WESTERN HEMISPHERE
WET AND DRY MAP OF WESTERN HEMISPHERE,
JANUARY 1, 1919

"Work is the curse of the drinking classes": In reversing the stern slogan of the Victorian moralists who proclaimed that booze was the bane of the proletariat, Oscar Wilde made a sarcastic comment on the decadence of the British upper classes—which included refined aesthetes such as himself.

Wilde's era, the late nineteenth century, had given rise to a Temperance movement, which had fought the excesses of alcohol (in the *lower* classes) by seeking to restrict its consumption—firstly by personal abstinence, but increasingly by campaigns for ever stricter legislation.

The call for Temperance, originating in the United States and Great Britain but not limited to Anglo-Saxon countries, evolved into a clamor for wholesale Prohibition: to make the production and consumption of alcohol illegal. Indeed, what better way to stamp out the excesses of drink but to forbid it entirely?

This was the objective of organizations such as the Anti-Saloon League, founded in Ohio in 1893 and quickly growing into a nationwide organization. It was the first one-issue political pressure group, campaigning for the prohibition of alcohol (its manufacture and sale) on county, state and federal levels. Its finest hour was the adoption of the 18th Amendment to the Constitution in 1920, establishing Prohibition throughout the entire United States. Success didn't last—Prohibition was repealed in 1933.

These maps date from the league's heyday, prior to the establishment of Prohibition. They show the success of its campaigns on a local level (black are "wet" areas, where alcohol may be served without restrictions, "gray" areas with some restrictions and "dry" areas in white):

The Wet and Dry Map of the Western Hemisphere shows the striking progress of the league's cause: In mid-1915, only the southern

WET AND DRY MAPS OF IDAHO, 1909, 1911, 1915, 1916

Jan. 1, 1909   Jan. 1, 1911

Jan. 1, 1915   Jan. 1, 1916

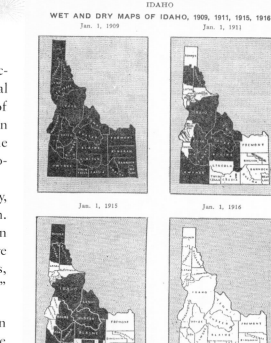
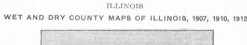

WET AND DRY COUNTY MAPS OF ILLINOIS, 1907, 1910, 1912

Jan. 1, 1912

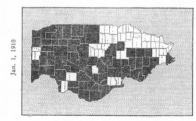
Jan. 1, 1910

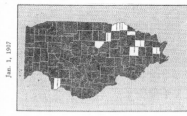
Jan. 1, 1907

"WET" AND "DRY" COUNTY MAP OF THE UNITED STATES
JANUARY 1, 1904

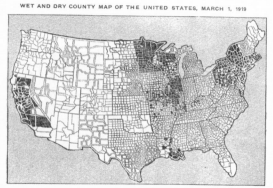
WET AND DRY COUNTY MAP OF THE UNITED STATES, MARCH 1, 1919

U.S. states (and two Canadian provinces) are predominantly dry—although there remain very few totally wet states: Alaska, Montana, Nevada, Pennsylvania. By New Year 1919, the whole of Canada and the United States is dry; strikingly, Prohibition was not at all a factor in Latin America.

The Wet and Dry County Map of the United States offers some nuances. At the start of 1904, only Maine, North Dakota and Kansas were totally dry, as were significant parts of Iowa, Oklahoma (the Indian Territory), Texas, Arkansas, Mississippi, Alabama, Tennessee, Kentucky, Georgia, Florida, North Carolina and the Virginias. The entire West and most of the Northeast remained wet. By March 1919, the situation had more than reversed: Apart from most of California, all of Minnesota, most of Michigan and Illinois, parts of Missouri and

Louisiana and much of the Northeast, the whole country was dry.

Two maps detail how the wet to dry evolution took place on a state level. Idaho was completely wet at the start of 1909, had started to dry up in the east and northwest by 1911, had suffered a reversal by 1915, with some counties reverting from dry to wet, but by 1916 had gone completely dry, probably due to the introduction of state-level legislation. Illinois shows a similar evolution, from almost completely wet to nearly all dry, with the exception of wet islands in and around Chicago and East St. Louis.

Although Prohibition has been repealed, its spirit (if I may use that word) is not entirely dead. The United States still has hundreds of dry counties, cities and towns, mainly in the South and Midwest.

7

Beautiful as a Snowflake: Fortress Coevorden

Its strategic location at a river crossing marked out the small Dutch town of Coevorden as a prime destination for tradesmen and warriors. Prosperity alternated with destruction. Coevorden was leveled in 1592, during the Dutch war of independence against the Spanish, and rebuilt to become Europe's most heavily fortified city of the seventeenth century. Economic stagnation helped preserve much of the fortifications to this day.

Fortress Coevorden's symmetrical layout is typical for what is known as a star fort. The Italian name of this type of fortress hints at the time of its origin. The *fortificazione moderna* evolved in mid-fifteenth-century Italy as a response to successful innovations in cannon technology. Gunpowder-propelled cannonballs rendered the medieval castle wall quite literally indefensible, even if the castle was high upon a hill.

In response, star forts were construed with flat, thick walls, consisting of triangle-shaped bastions, protected by moats and jutting outward to cover as much of the enemy battle-field as possible with firepower. Over the next three centuries, construction of the forts was perfected to a science throughout Europe.

Star forts changed the nature of European warfare, which became less battlefield-oriented and more siege-prone. Michelangelo tried his hand at star forts and his Italian compatriots were much in demand throughout the Continent for their fort-constructing skills. Taking the intricacies of star forts to their baroque extreme were the prolific Frenchman Vauban and Menno van Coehoorn, the Dutch soldier-engineer.

Fortress building was further elevated from a science to an art form, with a delicious vocabulary containing elegant terms such as ravelins and redoubts, scarps and counter-scarps, lunettes and bonnettes. The symmetry that started as an accident of military expedience developed into a thing of beauty, as intricately lovely as a snowflake.

And what a snowflake Fortress Coevorden is. A snowflake in full body armor, with several polygons' worth of walls and ditches

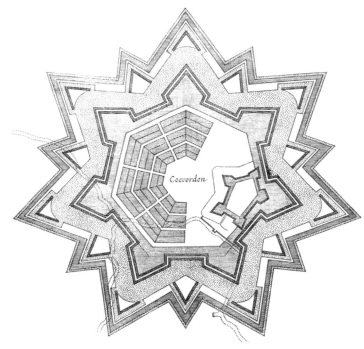

Coevorden

protecting the inner core of the fort. They don't make 'em like that anymore. And for a reason as practical as the star forts' three-centuries-long lucky streak—technology had moved on. The invention of explosive-shell munitions in the nineteenth century ended their military usefulness.

8

National Dishes of Brazil

Brazil conjures up conflicting images, of rapid economic development and deeply entrenched poverty, of vast rainforests and huge coastal cities. Brazil covers half of South America, both territorially and population-wise, and is its premier economic power but also the odd one out—the only Portuguese-speaking nation on a continent where Spanish dominates in almost every other country.

Film director Terry Gilliam named his 1985 black comedy *Brazil* because its Orwellian tone and depressing atmosphere (literally—it never stops raining) contrasted so starkly with the tropical, festive echo of that name. But even before Gilliam's movie, Brazil had a sinister connotation, as one of the Nazis' favorite hiding places post-1945. Think *The Boys from Brazil* (the book and the movie). Or the military dictatorship that ruled Brazil for much of the 1960s and 1970s, reveling in the slogan *Brasil, ame-o ou deixe-o* ("Brazil, Love It or Leave It").

That slogan is paraphrased here to describe one of Brazil's top exports: *Bifão, ame-o ou deixe-o* ("Beef, Love It or Leave It"). As with many things Brazilian, the country's beef production is both huge and hugely controversial. In the first years of the twenty-first century, Brazil overtook Australia as the world's leading beef exporter. But large herds of cattle are increasingly seen as major contributors to environmental problems as wide-ranging as the destruction of the rainforest (to graze your future hamburger) and the emission of green-house gases (i.e., the accumulated farts of all those cows).

The second of these two "national dishes" of Brazil carries a more neutral title—*A Cara do Brasil* ("The Face of Brazil")—and is a collection of typical Brazilian foods. This nice bit of culinary cartography is of course only possible if the country in question has a distinct shape. Brazil is fortunate to have an instantly recognizable shape that is easily able to accommodate the national dishes of the country. A place like Kansas would be too rectangular to represent anything but flapjacks, for example, and Chile would work only if the national dish was eels, spaghetti or some other food-stuff typified by its elongated shape.

9

Great Britain as a Cloud

It's a warm, windy summer's day. Woolly tufts of cloud drift by, punctuating the heat of the sun's rays, molded into swiftly shifting shapes by the wind that carries them. This is one of those days in which the cloudscape is one of nature's most glorious spectacles (instead of the leaden seal by which it usually obstructs a view of the heavens). What a great day for cloudspotting!

Cloudspotters who take their hobby seriously—oh yes, they do exist—will be able to tell you whether a cloud is a stratocumulus or a cirrustratus, whether it is calvus, capillatus or any other subspecies. Children, the other category of avid cloudspotters, will simply observe, "Look, there's a dog! And that's a dragon! And there goes Auntie Emma!"

Somewhere between the scholarly and the childlike is the cartophiliac cloudspotter—he or she (but let's face it, usually a *he*)

who scours the skies for clouds that look like countries. Or, if he is especially knowledgeable and desperate, clouds that look like provinces, counties and other subdivisions. It's a very discouraging hobby, as often you need to stretch your imagination to the breaking point in order to see a country-shaped cloud.

But sometimes you *do* get lucky—and what a fantastic moment that must be! We can only guess how Rob Gandy felt when he saw this cloud slowly transform before his very eyes into his own country—and above his own country too! "It had been more *solid* before I managed to get my camera," Gandy says of this UK-shaped cloud he snapped near Wadebridge in Cornwall on the morning of August 3, 1996. "As I watched, it slowly but surely broke up. Perhaps it was a portent of the effects of devolution following Tony Blair's election victory the following year."

The cloud formation does seem to give a relatively accurate—and quite symbolic—representation of Great Britain: The main bulk of the cloud is in the south, where most of the people live. Scotland's lower density sprouts neatly at the top, in a remarkably accurate position. Equally recognizable are East Anglia's bulge into the North Sea (in the east, obviously), the slower jut of Kent into the Channel (just south of East Anglia), the protrusion of Cornwall into the Atlantic, and England's slightly tilted southern coast between them. Wales is also recognizable, but at its northwest is missing Holyhead, which gives it its squareish look in real life. Northern Ireland is too big and too close to the "mainland," though, but all in all, not bad going for a nice morning cloud in August.

ACKNOWLEDGMENTS

Many thanks to all the cartophiles who visit, comment on and contribute to the Strange Maps blog, helping it grow. A big thank-you to all who were kind enough to contribute maps to this atlas. Special thanks to Joel Winten and Jos Gorssen for their frank and helpful assessment of the original manuscript; and to Jeff Galas, Megan Newman and Miriam Rich at Viking Studio, for believing in the book. And thank heavens for strong, black coffee, which played an instrumental role in getting this done.

NOTES

INTRODUCTION: NOT FOR NAVIGATION

xi *an anti-atlas of sorts (geography buffs might appreciate the double entendre)* The first book-bound map collection to be called by its modern name was Gerardus Mercator's *Atlas, Sive Cosmographicae Meditationes de Fabrica Mundi* (1595). Mercator named his book after the mythical Mauritanian king Atlas, reputed to have produced the world's first globe and often confused with the Greek mythological figure Atlas, a Titan who was condemned to carry the weight of the heavens on his shoulders. Predictably, confusion ensued. And depictions of Atlas upholding the celestial sphere evolved into him carrying the earth instead. Atlas is also a North African mountain range, its name possibly related to that of the aforementioned Mauritanian king, who must have ruled these parts. The Atlas chain is flanked by another, called the Anti-Atlas.

II. LITERARY CREATIONS

1. More's Memento Mori: A Color Map of Utopia

10 *"A map of the world that does not include Utopia is not worth even glancing at," said Oscar Wilde* In his 1891 essay "The Soul of Man Under Socialism."

10 *the perfect society of Utopia* "Utopia" is a Greek neologism that can be translated as "No Place."

10 *a traveler called Raphael Hythlodaeus* A made-up Greek surname that can be rendered as "Nonsense-Dispenser."

10 *its main city Amaurotum* Greek for "Mist Town."

10 *the river Anydrus* Greek for "waterless."

10 *in the form of a memento mori* Latin for "Remember that you shall die."

3. Finally in Need of a Navy: New Switzerland

14 *Jules Verne published* Seconde patrie *(Second Fatherland) in two parts* First English publication as *Their Island Home* and *The Castaways of the Flag*, later published in one volume as *The Castaways of the Flag*.

15 *Check out these pairs* If only one name is given, that name is used on both the German and French maps.

4. Lord of the Flies, with a Happy Ending: *A Two Years' Vacation*

17 *the setting for* Deux ans de vacances *(1888)* Published as *A Two Years' Vacation* in the United States (two volumes) and *Adrift in the Pacific* in the United Kingdom (one volume) the following year. It was republished in 1965 in two volumes as *Adrift in the Pacific* and *Second Year Ashore*. A new one-volume translation came out in 1967 as *A Long Vacation*.

III. ARTOGRAPHY

2. Drawn from Memory: United Shapes of America

26 *also represented quite frequently is Texas's southern extremity* Some commenters suggested this might also represent Louisiana's Mississippi Delta.

26 *The Great Lakes' indentation in the U.S.'s northern border is quite difficult to render from memory (give it a try!) and therefore generally ignored or minimized* True, Lake Michigan is wholly within the United States, but even allowing for the fact that water surfaces might be included in the "territory" of a country, the four other Great Lakes are shared with Canada and should be represented by a marked concavity of the border.

4. Now This Is World Music: Harmonious World Beat

30 *the result doesn't sound half bad* Go check it out for yourself at http://limulus.net/files/misc/world.mp3.

IV. ZOOMORPHIC MAPS

1. Scott's Great Snake; or, The Anaconda Plan

42 *Maryland is labeled "We give in"* Possibly a reference to the fact that Maryland, although a slave state, stayed in the Union.

42 *crowned with a phrygian cap* A conical cap worn in antiquity by the inhabitants of Phrygia (central Asia Minor); to the Greeks, it symbolized oriental barbarism, to the Romans, it stood for liberty (as it was worn by freedmen during festivals). A phrygian cap atop a Liberty Pole came to symbolize freedom during the American Revolution, a symbol that was adopted and spread by the French Revolution. More recently the Smurfs have been among its most enthusiastic adepts.

V. (POLITICAL) PARODY

2. The Jesusland Map

53 *and even inspired a pop song by the same name* "Jesusland," off Ben Folds's 2005 album *Songs for Silverman*, reached number 42 in the UK airplay charts on October 2 of that year.

VII. OBSCURE PROPOSALS

8. Sweet Home Talladego: C. Etzel Pearcy's Thirty-eight-State Union

81 *with crooked borders that dissect all but six states?* Maine, Vermont, Rhode Island, Delaware, Louisiana and Hawaii.

82 *now cut through large metropolitan areas* For example., Kansas City straddling Kansas and Missouri; New York City covering New York, New Jersey and Connecticut; and Washington, D.C., spilling over into Maryland and Virginia.

82 *and mold each new state around one metropolis* For example, Chicago at the center of the state of Dearborn, Atlanta in the middle of the state of Piedmont, St. Louis at the heart of the state of Osage.

9. Forty-seven Fingers in the Antarctic Pie: The Frontage Principle

83 *you could visit six of those seven countries* The Norwegian claim is only defined east-west, not north-south, and therefore in theory does not extend all the way down to the pole itself.

84 *Fourteen from the Americas* Greenland, though an autonomous part of Denmark, is habitually counted as part of the American continent.

10. Out of One, Many: Sixteen New American Nations

85 *The unofficial motto of the United States is* E Pluribus Unum The official one, since 1956, is "In God We Trust"; an act of Congress in 1782 has codified *E Pluribus Unum* as one of the mottoes on the Great Seal of the United States.

85 *and proposes to reverse the phrase* The correct Latin translation would be *Ex Uno, Plures*.

87 *"Only in theory will Olympia's governmental powers reach past the Sierra Nevada"* This should be the Cascades; the Sierra Nevada is farther south, mainly in California.

VIII. EPHEMERAL STATES

3. A Small Country That Never Was: Greater Belgium

94 *Julius Caesar mentions the Belgae as the "bravest of all Gauls"* The Belgae were a collection of Celtic and Germanic tribes occupying present-day Belgium, plus parts of northern France and southern England. Mention of their bravery in the opening lines of his *De bello Gallico* is a fact related with much pride by the Belgians, conveniently forgetting that Caesar meant by this that they were the least touched by Roman civilization.

5. A Pizza Slice Called Friendship: The World's First Esperanto State

98 *This* Drielandenpunt *once was a* Vierlandenpunt *("quadrinational point")—arguably the only one in the world ever* One could argue that Moresnet never really was a country, and that therefore only three countries met at the quadripoint, plus the condominium of two of them. Moreover, there is some discussion whether the borders of Zambia, Zimbabwe, Botswana and Namibia "touch" in the Zambezi River; most sources disagree. A well-known quadripoint between U.S. states is the Four Corners, where Utah, Arizona, New Mexico and Colorado touch.

IX. STRANGE BORDERS

1. The Circle and the Wedge: Delaware's Curious Border

102 *The Twelve-Mile Circle, the border with Pennsylvania and the only circular boundary between U.S. states* Unless you consider the many straight-line borders based on latitude (and therefore centered on the North Pole) as part of an arc. Which you could.

102 *the entire Delmarva Peninsula* The name of this peninsula is composed of the three states occupying it—DELaware, MARyland, VirginiA. The Native Americans called it Accomac, but it's unclear whether it ever had a "proper" name since. Oh, and to add to the confusion, Delmarva isn't a peninsula anymore since the Chesapeake and Delaware Canal cut it off from the mainland in the nineteenth century.

3. The Aroostook War (Bloodless if You Don't Count the Pig)

106 *Surveyors sent out in 1820 by Maine to mark out the new state's territory* Until then it had been the District of Maine, a noncontiguous part of Massachusetts.

106 *This frontier version of a Sitzkrieg* "Sitting war," after the period between the declaration of war between France and Germany and actual combat at the beginning of World War II.

106 *became known as the Aroostook War* Other names include the Pork and Beans War and the Lumberjack War.

4. World's Smelliest Border: The Limburg Split

108 *In one of Mark Twain's stories* "The Invalid's Story" (1877).

108 *the Belgian city of Limbourg (formerly the capital of an eponymous duchy)* The stinky cheese is named after this duchy.

108 *Those two provinces formed a patchwork of allegiances to dukes, counts and prince-bishops until they were unified into one administrative region by the conquering French in 1795* Although most of Belgian Limburg once formed the County of Loon.

6. May the Sauce Be with You: Battle Lines in the Barbecue Wars

112 *"Tell me what you eat, and I'll tell you who you are": The famous quote by legendary gastronome Brillat-Savarin*   Jean Anthelme Brillat-Savarin (1755–1826) was, improbably, first a deputy to the National Assembly during the French Revolution, then first violin in New York's Park Theater, and finally a judge at the Court of Cassation in Paris. Only two months before his death did he publish *The Physiology of Taste*, one of history's most influential books on gastronomy. Another one of his food quotes: "Dinner without cheese is like a beautiful woman with only one eye."

X. EXCLAVES AND ENCLAVES

4. Bubbleland, Not Far from Monkey's Eyebrow: The Kentucky Bend

120 *and even "Bubbleland"—quite an image-provoking epithet; one involuntarily pictures Michael Jackson's monkey's own version of Neverland*   Of course so named because of its shape, although the nearby Kentucky hamlet of Monkey's Eyebrow leaves some wriggle-room for other explanations.

6. Cooch Behar: The Mother of All Enclave Complexes

123 *this map by Dr. Brendan Whyte, who wrote a thesis on the enclaves, counterenclaves and counter-counterenclave (the world's only one!) of Cooch Behar*   Dr. Brendan Whyte, *Waiting for the Esquimo: An Historical and Documentary Study of the Cooch Behar Enclaves of India and Bangladesh*, School of Anthropology, Geography and Environmental Studies, University of Melbourne (2002).

125 *Brecht used the name of the (then still) princely state to symbolize the far end of the world*   Another reference is in the title song of the 1968 Julie Andrews film *Star* (which of course rhymes with Cooch Behar), with the same implication of exotic distance.

XI. A MATTER OF PERSPECTIVE

7. The Eclectic Archipelago and Lakes of All Countries

141 *New Zealand's "South" Island is called Stewart Island today*   In the nineteenth century, New Zealand was often described as having a North Island (still called so today), a Middle Island (then also called New Munster, today called South Island) and a South Island (then also called New Leinster, today Stewart Island).

XIII. LINGUISTIC CARTOGRAPHY

1. The World's Linguistic Superpower: Papua New Guinea

154 *The* Ethnologue's *most recent information lists the following ten countries as containing the largest number of living languages (indigenous and imported), corresponding with the countries on Mr. Parkvall's map*   Counting languages is an inexact discipline: What one linguist considers a dialect might seem a separate language to the next one. Estimates for the number of languages remaining on earth range from 6,000 to 12,000. The *Ethnologue* lists 6,912 living languages on the planet.

XIV. BASED ON THE UNDERGROUND

2. Itineraries into Eternity (and Back, Occasionally)

164 *Line 3a, throwing unbaptized babies straight into Limbo, was "closed" by Pope Benedict XVI in 2006*   The exclusion from Paradise of unbaptized children, too young to have committed personal sins but still affected by original sin, "is in contradiction to Jesus Christ's special affection for the little ones," according to the International Theological Commission of the Vatican, discounting a theological point of view that was previously "permissible" to hold.

4. A Diagram of the Eisenhower Interstate System

168 *The Dwight D. Eisenhower National System of Interstate and Defense Highways (Eisenhower Interstate System for short, or EIS for even shorter) spans the entire United States, including Alaska and Hawaii*   "Interstate" refers to the fact that these highways were funded federally rather than that they cross state lines.

168 *It serves all major American cities*   The largest city in the contiguous United States not served by an interstate is Fresno, California.

XV. FANTASTIC MAPS

2. Jamerica the Beautiful

174 *Map nuts similarly observe pareidolia of states and continents in everyday objects*   From the Greek *para* ("beside") and *eidolon* ("image"), a phenomenon whereby a random stimulus is perceived as meaningful.

3. One Ring to Rule Them All, Mate: Tolkien's Australia

176 *Previously named Spring Pound, this geographic feature in Australia's Red Center*   In Australian English, a pound can refer to a deep valley enclosed by cliffs.

XVI. CARTOGRAMS AND OTHER DATA MAPS

2. High Noon in Washington, D.C.: Keeping Time Before Trains

183 *No Hillary, it's not 3 a.m. in Washington—it's noon*   In fact, it could be either noon or midnight, but comparison with the time in other localities leads one to the conclusion that it must be noon in D.C.

4. If Wikipedia Was China, Then . . .

186 *"In the hierarchy of users, the vast majority of visitors to Wikipedia, 217 million of them, are readers"*   Mr. San Miguel calculated estimates for mid-2007 Wikipedia contributions. Regular: 217M/338K = 642:1; Active: 217M/105K = 2,055:1; Very Active: 217M/14K = 15,585:1

5. Ulstermen, Tennesseans, Mexicans and Germans: A Tale of Texas's Squares

187 *One of Northern Ireland's many underappreciated contributions to the United States is the central public square as an element of urban planning*   Another is the number of presidents of Ulster stock—varying from eleven to sixteen, depending on how you count. But even the lowest figure accounts for a quarter of all U.S. presidents.

187 *Only one Texas county originally had a Marshall-
type courthouse square* Harrison County, on the
Louisiana border.

188 *The Four-Block courthouse square is almost as
rare, occurring in only three counties* Schleicher,
Parker, and Rusk (west to east).

**9. Even Penguins Can Go Online: Country Code Top-
Level Domains**

197 *.gg (Guernsey): gaming and gambling industry,
particularly in relation to horse racing ("gee-
gee")* In British English, "the gee-gees"
describes horses, especially in a racing context;
the word derives from "gee," a command for the
horse to turn right.

XVII. MAPS FROM OUTER SPACE

1. A Foldable Map of Mars's Moon Phobos

202 *Phobos is the larger and closer of these
two* Both discovered on August 18, 1877, by
American astronomer Asaph Hall Sr.

202 *Contributing to the moon's irregular shape is
Stickney Crater* Named after Angeline Stickney

Hall, the discoverer's wife. Many other features
on Phobos are named for characters or places in
Jonathan Swift's *Gulliver's Travels* (1726) (e.g.,
Drunlo, Flimnap, Clustril, Limtoc and Reldresal)
in that book, the inhabitants of the floating island
of Laputa discovered . . . two moons near Mars.

202 *Before spacecraft sent images back, showing
Phobos to be a natural object* Starting with
Mariner 9 in 1971, most recently by Mars Express
in 2004.

3. One Small Stroll for Man: The First Moonwalk

205 *After traveling hundreds of thousands of
miles, the landing crew of the Apollo 11 lunar
mission spent two and a half hours on the lunar
surface* On average, the distance between the
earth and the moon is 238, 857 miles (384,403
km), core to core. This equals about thirty times
Earth's diameter.

205 *with Armstrong making just one dash at the
other side's goal* Football fans (i.e., soccer
aficionados), discuss: Is Armstrong offside?

XVIII. WATCHAMACALLIT

4. Clouds with Silver Linings: Europe Discovers the World

218 *Portuguese seafarers had rounded the Cape of
Good Hope in 1488 and found a trade route to
India* Aptly named the "Cape of Storms" by
Bartolomeu Dias in 1488, later given a name by the
Portuguese king reflecting its positive impact on
trade.

218 *at the death of the emperor Charles V in
1558* Not 1551, as indicated above the map.

7. Beautiful as a Snowflake: Fortress Coevorden

223 *Its strategic location at a river crossing marked
out the small Dutch town of Coevorden
as a prime destination for tradesmen and
warriors* The Dutch term "Coevorden" refers
to the spot where farmers forded their cows
across the river and can thus be translated as
"Cowford," a female pendant to Oxford, as it
were. Incidentally: The English descendant of
a Dutchman surnamed Van Coevorden gave
his—slightly abbreviated—name to the Canadian
city of Vancouver.

IMAGE CREDITS

I. CARTOGRAPHIC MISCONCEPTIONS

1. Nice Colors, Wrong Height: Mapping Mountains in 1831
Anthony Finley, Table of the Comparative Heights of the Principal Mountains &c in the World (1831).
Image courtesy of the David Rumsey Map Collection.
www.davidrumsey.com

2. Black Rock, Fake Rock: The First, False Map of the True North
Gerardus Mercator, Septentrionalium Terrarum Descriptio (ca. 1595).
Wikimedia Commons image of map in the public domain.

3. Black Amazon Women: The Island of California
Johannes Vingboons, Map of California as an Island (ca. 1650).
Wikimedia Commons image of map in the public domain.

4. "We Hope the League of Nations Will Rule the Tetrahedron Well"
What the World May Come To: Map of the Earth as a Tetrahedron, from *My Magazine* (May 1918)
Image courtesy of Marcus L. Rowland.
http://homepage.ntlworld.com/forgottenfutures/

II. LITERARY CREATIONS

1. More's Memento Mori: A Color Map of Utopia
Ambrosius Holbein, Utopia (1518).
Image courtesy of Prof. Em. Ulrich Harsch of the Bibliotheca Augustana, Fachhochschule Augsburg.
http://www.hs-augsburg.de/~Harsch/augusta.html

2. Love's Topography: La Carte de Tendre
François Chauveau, La Carte de Tendre (1654).
Wikimedia Commons image of map in the public domain.

3. Finally in Need of a Navy: New Switzerland
Originals are free of copyright by international regulations, copyright for this reproduction granted by Bernhard Krauth and the Jules Verne Club of Germany.
www.jules-verne-club.de
www.jules-verne.eu

4. *Lord of the Flies*, with a Happy Ending: *A Two Years' Vacation*
Originals are free of copyright by international regulations, copyright for this reproduction granted by Bernhard Krauth and the Jules Verne Club of Germany.
www.jules-verne-club.de
www.jules-verne.eu

5. Captain Nemo's Deathbed: The Mysterious Island
Originals are free of copyright by international regulations, copyright for this reproduction granted to the publisher and by Bernhard Krauth and the German-language Jules Verne Club.
www.jules-verne-club.de
www.jules-verne.eu

6. Not Kansas, but Just as Square: The Land of Oz
The Land of Oz.
Original publication © 1914 L. Frank Baum.
Image courtesy of Eric and Laura Gjovaag, The Wonderful Wizard of Oz Web site.
www.thewizardofoz.info

7. Close Your Eyes and Think of Airstrip One: The World in 1984
A Map of the World in "Nineteen Eighty-Four."
Wikimedia Commons image released into the public domain under the CC-BY-SA-2.5 license.

III. ARTOGRAPHY

1. Back to the Drawing Board: An Inaccurate Map of Charlottesville
Russell Richards, *An Inaccurate Map of Charlottesville*, lithograph.
© Russell Richards, image reproduced with permission.
www.russellrichards.com

2. Drawn from Memory: United Shapes of America

Kim Dingle, *United Shapes of America (Maps Drawn by Las Vegas Teenagers)* (1991), oil on wood, 48 × 72 inches. Private collection.

Image reproduced courtesy of the artist and Sperone Westwater, New York.

http://www.speronewestwater.com

3. Gothic Barcelona: Horror and Humor in El Born

Image from *A Weird and Wonderful Guide to Barcelona*, Le Cool Publishing.

© Vasava Artworks for Le Cool Publishing.

www.lecool.com/books

4. Now This Is World Music: Harmonious World Beat

© 1996 James Plakovic Music*Art*. Visual & Music Copyrights apply. Reproduction without permission is strictly prohibited. All rights reserved. Unauthorized duplication is a violation of applicable laws.

http://plakovic.com/

5. Chimero's Chimeras: Transformed States of America

Frank Chimero, *The States: California*.
© 2008 Frank Chimero.
www.frankchimero.com

Frank Chimero, *The States: Illinois*.
© 2008 Frank Chimero.
www.frankchimero.com

6. Pretentious, Moo? The Cow That Conquered the World

© 2008 TNT Post. Reproduced with permission.
Image courtesy of Diedrik A. Nelson.
www.tntpost.nl
http://www.danstopicals.com

7. Don't Do It, Baby: Giant Infant Threatens Downtown Vancouver

Aaron Meshon, *Baby in Vancouver Harbour* (published in *Vancouver Magazine*, 2005).
© Aaron Meshon.
www.aaronmeshon.com

8. "Not Atoll": Real Maps Reassembled into Imaginary Places

"Not Atoll" © Francesca Berrini 2008.

Mixed Media on Canvas.
www.wolfecontemporary.com
www.gibsongallery.com

9. Maps as Art: Painted Peru

Christa Dichgans: *Peru* (2004).
Image courtesy of Contemporary Fine Arts Galerie, Berlin. Photo: Jochen Littkemann.
www.cfa-berlin.com

IV. ZOOMORPHIC MAPS

1. Scott's Great Snake; or, The Anaconda Plan

J. B. Elliott, *Scott's Great Snake* (Cincinnati, 1861).
Image from the Library of Congress Geography and Map Division, Washington, D.C.; image in the public domain.

2. Itching to Spread Its Wings West: The American Dove, 1833

Joseph & James Churchman, I. W. Moore, The Eagle Map of the United States Engraved for Rudiments of National Knowledge (1833).
Image courtesy of the David Rumsey Map Collection.
www.davidrumsey.com

3. Geographical Fun: The Aleph Maps

"Aleph" (William Harvey), *Geographical Fun: Being Humourous Outlines of Various Countries, with an Introduction and Descriptive Lines* (London, ca. 1868).
Image from the Library of Congress Geography and Map Division; image in the public domain.

V. (POLITICAL) PARODY

1. A "New Species of Monster": The Gerry-mander

Boston Gazette, The Gerry-mander: A New Species of Monster, Which Appeared in Essex South District in Jan. 1812 (Salem, 1812).
Image from the Library of Congress Printed Ephemera Collection; image in the public domain.

2. The Jesusland Map

Gavin Webb, Jesusland (2004).
© Gavin Webb 2004.

3. Flanders Sleeps with the Fishes: Wallonie-sur-Mer

© 2006 Le Soir, Mortierbrigade.
www.lesoir.be
www.mortierbrigade.com

4. A New Simplified Map of London

© Ellis Nadler.
www.nadler.co.uk

VI. MAPS AS PROPAGANDA

1. Overlapping Claims: Early-Twentieth-Century Balkan Aspirations

Report of the International Commission to Inquire into the Causes and Conduct of the Balkan Wars, Map of Balkan Aspirations, showing boundaries of 1912 (1914).
Image courtesy of the Perry-Castañeda Map Collection at the University of Texas at Austin.

2. Germany Wins World War I: French Worst-Case Scenario

Map courtesy of Anthony Langley.
www.greatwardifferent.com

3. Another Plan to Kill France: Italy on the Atlantic

The British Dominions Year Book, A Pan-German Scheme for the Extinction of France (1918).
Image courtesy of the Perry-Castañeda Map Collection of the University of Texas at Austin.

4. A Patchwork Germany to Keep France Safe

The British Dominions Year Book: The Reconstruction of Europe (1918).
Image courtesy of the Perry-Castañeda Map Collection of the University of Texas at Austin.

5. Germany's Future: From the English Channel to the Caspian Sea

The British Dominions Year Book, Germany's Future (1918).
Image courtesy of the Perry-Castañeda Map Collection of the University of Texas at Austin.

6. What a German Africa Might Have Looked Like

The British Dominions Yearbook, German Claims in Africa 1917 (1918).
Image courtesy of the Perry-Castañeda Map Collection of the University of Texas at Austin.

7. Reverse-Engineered Nazi Propaganda: "Germany Must Perish"
© Theodore N. Kaufman, image kindly provided by Randall L. Bytwerk of Calvin College, Grand Rapids (MI).
http://www.calvin.edu/cas/gpa/

VII. OBSCURE PROPOSALS

1. Dampieria to Guelphia: A Ten-State Australia
Journal of the Royal Geographical Society, Volume 8, Australia, according to the Proposed divisions (London, 1838).
Image courtesy of the Perry-Castañeda Map Collection of the University of Texas at Austin.

2. "That Absurd Element in Jefferson's Mind": Ten States That Never Were
The Northwest Territory Celebration Committee, Historical Map of the Old Northwest Territory (1937).
Image courtesy of John A. Lindquist.

3. Texas Can't Hold 'Em: The Lone Star Empire
© Donald S. Frazier, Ph.D., Abilene, Texas.

4. Western Designs on the Middle East: The Sykes-Picot Agreement
Map of the 1916 Sykes-Picot Agreement.
Reproduced with permission from the Palestinian Academic Society of the Study of International Affairs.
www.passia.org

5. Franz Ferdinand's Winning Idea: The United States of Greater Austria
© Thomas Griesbacher.

6. Oklahoma's Stillborn Twin: The State of Sequoyah
D. W. Bolich, State of Sequoyah map, compiled from the USGS Map of Indian Territory (1902), revised to include the county divisions made under direction of Sequoyah Statehood Convention (1905).
Image from the McCasland Digital Collection of Early Oklahoma and Indian Territory Maps, reproduced with permission of the Oklahoma State University Edmon Low Library.
http://okmaps.library.okstate.edu/

7. Dutch Dreams of Expansion into Germany: "Eastland, Our Land"
Image courtesy of Tubantia, available from Wikipedia under the terms of the GNU Free Documentation license version (1.2 or later) published by the Free Software Foundation.
http://de.wikipedia.org/wiki/Bild:Bakker_Schut-plan.PNG

8. Sweet Home Talladego: C. Etzel Pearcy's Thirty-eight-State Union
Image reproduced with permission of Plycon Press.

9. Forty-seven Fingers in the Antarctic Pie: The Frontage Principle
Image by Paul Youlten, reproduced under cc-by-sa 3.0 license.

10. Out of One, Many: Sixteen New American Nations
© 2004 Matt Kirkland; Divide and Conquer.
www.exunumpluribus.com

11. The Fro Gymraeg: A Reservation for Welsh
Map reproduced with kind permission of Aran Jones, Cymuned.
http://cymuned.net/

VIII. EPHEMERAL STATES

1. The Darién Scheme: A Tropical Scotland Gone Horribly Wrong
Herman Moll, Scots Settlement in America Called New Caledonia, A.D. 1699 (1736).
Image courtesy of the David Rumsey Map Collection.
www.davidrumsey.com

2. Gray Area Between the United States and Canada: The Republic of Indian Stream (1832–35)
Wikipedia image added by Citynoise, later versions by AnonMoos, distributed under the Creative Commons Attribution ShareAlike 2.5 license (CC-BY-SA-2.5).

3. A Small Country That Never Was: Greater Belgium
Jeremiah Greenleaf, Belgium and Holland (1840).
Image courtesy of the David Rumsey Map Collection.
www.davidrumsey.com

4. Nice Name for It: The United States of Stellaland
Scottish Geographical Magazine, Sketch Map of South Africa Showing British Possessions, July 1885.
Image courtesy of Lindsay F. Braun.

5. A Pizza Slice Called Friendship: The World's First Esperanto State
Wikipedia image by Ed Stevenhagen, released into the public domain.

6. Independent for Only Twenty-four Hours: Carpatho-Ukraine
Wikipedia map by PM, released into the public domain.

IX. STRANGE BORDERS

1. The Circle and the Wedge: Delaware's Curious Border
Map created for Wikipedia by Eoghanacht, modified by Lasunncty, released under the GNU Free Documentation License.

2. Going, Going, Gone: The Old Cherokee Country
James Mooney, The Cherokee Country, from the 19th Annual Report of the Bureau of American Ethnology (1900).
Image courtesy of the Perry-Castañeda Map Collection of the University of Texas at Austin.

3. The Aroostook War (Bloodless if You Don't Count the Pig)
Maine Boundary Controversy, 1782–1842, published in *The American Nation: A History from Original Sources* (New York: Harper & Brothers, 1906).
Image in the public domain, kindly provided by Chip Gagnon.
http://www.upperstjohn.com/aroostook/hartmap.htm

4. World's Smelliest Border: The Limburg Split
© 2008 TNT Post. Reproduced with kind permission.
Image courtesy of Diedrik A. Nelson.
www.tntpost.nl
www.danstopicals.com

5. An Accident of Nature, Geography and Man: Market Reef's Jigsaw Border
Map drawn by the author, based on a map at www.histdoc.net, with kind permission of Pauli Kruhse.

6. May the Sauce Be with You: Battle Lines in the Barbecue Wars
Barbecue Map of South Carolina, from *South Carolina: A Geography* by Charles F. Kovacik and John J. Winberry. Reproduced with permission.

X. EXCLAVES AND ENCLAVES
1. You Say Enclave, I Say Exclave: Let's Fence the Whole Thing Off
© Jan S. Krogh's Geosite (Vilnius, Lithuania) http://geosite.jankrogh.com.

2. The Vennbahn Complex: Five Little Germanies in Belgium
© 2008 Patrik Fagard.

3. Enclaves, Counterenclaves and a Dead Body: The Borders of Baarle
© 2008 Patrik Fagard.

4. Bubbleland, Not Far from Monkey's Eyebrow: The Kentucky Bend
Map at Wikipedia by Jim Efaw, licensed under the Creative Commons Attribution ShareAlike versions 2.5, 2.0 and 1.0 (CC-By-SA).

5. Madha and Nahwa: The Omelet-Shaped Enclave Complex
© National Geographic Society. Reproduced with kind permission.

6. Cooch Behar: The Mother of All Enclave Complexes
The Cooch Behar Enclaves Illustration by Brendan Whyte.

XI. A MATTER OF PERSPECTIVE
1. Subcontinent with a Twist: Sri Lanka on Top!
© Himal Southasian Magazine.
www.himalmag.com

2. Your Antipodes Most Likely Have Fins
Wikipedia map of antipodes of the Earth, in Lambert Azimuthal Equal-Area Projection. Licensed under the Creative Commons Attribution-ShareAlike 2.5 License.

3. Go with the Flow: The Norwegian Drop
Den Norske Dråpen, © Rolf Groven. Image reproduced with kind permission.
http://www.groven.no/rolf/

4. Humboldt's Imaginary Mountain: The Beginning of Geobotany
Joseph Meyer, Botanische Geographie. Humboldts Pflanzenregionen in aequinoctial America, nach ihrer Erhebung über die Meeresflaeche (1860).
Image courtesy of the David Rumsey Map Collection.
www.davidrumsey.com

5. One Size Fits All: The Equinational Projection
© 1994 Reeves, C. IASBS Equinational Projection. *Globehead! Journal of Extreme Geography*, Volume One, Thing 1, pp. 18–19.

6. Courses, Countries and Comparative Lengths: The World's Principal Rivers
Society for the Diffusion of Useful Knowledge, A Map of the Principal Rivers Shewing Their Courses, Countries and Comparative Lengths (1834).
Image courtesy of the David Rumsey Map Collection.
www.davidrumsey.com

7. The Eclectic Archipelago and Lakes of All Countries
G. W. Colton: *Comparative Size of Lakes and Islands* (1885).
Image courtesy of the David Rumsey Map Collection.
www.davidrumsey.com

8. Some Like It Hot—and Wet: The Luscious Waterworld of Dubia
All Dubia images © 2008 Chris Wayan.
www.worlddreambank.org/d/dubia.htm

XII. ICONIC MANHATTAN
1. "No One Calls Clinton Clinton": Pinning Down Manhattan's Neighborhoods
Neighborhoods of Manhattan.
© 2006 Alexander R. W. Cheek.

2. From Spuyten Duyvil to Battery Park: A Wordmap of Manhattan
© 1997 Howard Horowitz.

Originally appeared in the *New York Times* on August 30, 1997.
www.wordmaps.net

3. A World Map of Manhattan
© 2005 Danielle Hartman.

4. Who Put the Ice in NYC?
© Aaron Meshon.
www.aaronmeshon.com

XIII. LINGUISTIC CARTOGRAPHY
1. The World's Linguistic Superpower: Papua New Guinea
© Mikael Parkvall.

2. Switzerland's Curious Culinary and Cultural Divide: The *Röstigraben*
Book cover of *Rideau de rösti*, *Röstigraben* by Laurent Flutsch, catalog of the exhibition "Rideau de rösti," Musée romain de Lausanne-Vidy and Infolio 2005.
Image courtesy of Laurent Flutsch.

3. Paris's Most Unexpected Export: The Dorsal /r/
© Mikael Parkvall.

4. Praise the Lord and Pass the Dictionary: Europe's Polyglot Prayers
Gottfried Hensel, Europa Polyglotta, Linguarum Genealogiam exhibens, una cum Literis, Scribendique modis, Omnium Gentium. From *Synopsis Universae Philologiae* (Nürnberg, 1741). Image in the public domain.

5. Before France Spoke French
© Mikael Parkvall.

XIV. BASED ON THE UNDERGROUND
1. Oslo to Pyongyang Without Changing Trains: World Tube Map
© Mark Ovenden.
From *Transit Maps of the world* (Penguin, 2007) by Mark Ovenden.

2. Itineraries into Eternity (and Back, Occasionally)
Eternal Traffic.
© 2007 Bernd Wagner. www.bwmedien.net

3. Cecil B.'s Niece and Other Musical Bee's Knees
The Tube Map of Musical Theatre History.
© John Howrey.

4. A Diagram of the Eisenhower Interstate System
A Diagram of the Eisenhower Interstate System.
© Rebecca C. Brown.

XV. FANTASTIC MAPS

1. The Whole World in a Cloverleaf
Heinrich Buenting, Die ganze Welt in einem
    Kleberblatt (1581).
Wikipedia map in the public domain.

2. Jamerica the Beautiful
Image courtesy of Bjørn A. Bojesen.

3. One Ring to Rule Them All, Mate: Tolkien's Australia
© James Hutchings.
www.ageoffable.net

4. Oh, Inverted World: If Land Was Sea and Sea Was Land
© Vladislav Gerasimov.
www.vladstudio.com

XVI. CARTOGRAMS AND OTHER DATA MAPS

1. Turn the Other Cheek: The French Kissing Map
© Gilles Debunne.
http://combiendebises.free.fr

2. High Noon in Washington, D.C.: Keeping Time Before Trains
A. J. Johnson, Diagram exhibiting the difference of
    time between the places shown & Washington
    (1860).
Image courtesy of the David Rumsey Map Collection.
www.davidrumsey.com

3. Vital Statistics of a Deadly Campaign: The Minard Map
Charles Minard, Carte figurative des pertes successives
    en hommes de l'armée française dans la campagne
    de Russie 1812–1813 (1869). Image via Wikipedia,
    map in the public domain.

4. If Wikipedia Was China, Then . . .
© 2008 Frank San Miguel.
www.tech4D.com/blog

5. Ulstermen, Tennesseans, Mexicans and Germans: A Tale of Texas's Squares
Reprinted with permission from *The Atlas of Texas*
    (Austin: Bureau of Business Research, University of
    Texas at Austin, 1976).

6. Super Interesting: The World from a Brazilian Perspective
All images courtesy of *Superinteresante* magazine, São
    Paulo.
http://super.abril.com.br/

7. Where's Australia, Where's Russia? A Cartogram of the World's Population
© 2006 SASI Group (University of Sheffield) and Mark
    Newman (University of Michigan).
www.worldmapper.org

8. Staring at the Sun: All Total Solar Eclipses Until 2020
Total and Annular Solar Eclipse Paths: 2001–2020.
    Eclipse predictions by Fred Espenak, NASA/
    Goddard Space Flight Center. Image courtesy of
    Mr. Espenak. For more information on solar and
    lunar eclipses, visit his eclipse Web site: sunearth
    .gsfc.nasa.gov/eclipse/eclipse.html

9. Even Penguins Can Go Online: Country Code Top-Level Domains
Designed by John Yunker of Byte Level Research,
    Country Codes of the World is a registered
    trademark.
www.bytelevel.com

10. Bubbles Bursting: Area and Population of Foreign Countries (1890)
Rand McNally, Area and Population of Foreign
    Countries, Compared with the United States
    (1890).
Image courtesy of the David Rumsey Map Collection.
www.davidrumsey.com

11. The Inglehart-Welzel Cultural Map of the World
© World Values Survey, reproduced with kind
    permission of Ronald Inglehart.

www.worldvaluessurvey.org

XVII. MAPS FROM OUTER SPACE

1. A Foldable Map of Mars's Moon Phobos
"A Foldable Map of Phobos, a Moon of Mars,"
    © 2008 Chuck Clark. Boundary parti: P. E.
    Clark; shaded relief: Astrogeology Team USGS;
    rectification: P. J. Stooke; control: P. C. Thomas;
    model: A. T. Öner.
http://rightbasicbuilding.com/
A review and a detailed assembly how-to can be found
    on this page of the Planetary Society: http://www
    .planetary.org/blog/article/00001348/

2. The Colorful Side of the Moon
The Central Far Side of the Moon (Scale 1:15.000.000),
    Image in the Public Domain. Courtesy Desiree E.
    Stuart-Alexander, USGS Astrogeology Research
    Program, http://astrogeology.usgs.gov

3. One Small Stroll for Man: The First Moonwalk
Apollo 11 Landing Site Map: Baseball Comparison.
    Image created by Thomas Schwagmeier, based on
    a suggestion by Eric Jones. Reproduced with kind
    permission.
http://history.nasa.gov/alsj/

Apollo 11 Landing Site Map: Football (Soccer)
    Comparison. Image created by Thomas
    Schwagmeier, based on a suggestion by Joseph
    O'Dea.
Reproduced with kind permission.
http://history.nasa.gov/alsj/

4. Naming Titan's Methane Sea
Map created by Peter Minton @ EVS Islands / 02-23-
    2008 from Source Image PIA 100008
    Courtesy of NASA/JPL-Caltech.
http://www.evs-islands.com/

XVIII. WATCHAMACALLIT

1. True or False: The Vinland Map
Wikipedia image, in the public domain.

2. Europe, if the Nazis Had Won: Neuropa
© Sampsa Rydman.
www.valtakunta.eu

3. Not Even Faux, Just Plain Wrong: Schoolcraft's Federation Islands
Henry Schoolcraft, Map of the Federation Islands
    (1820). With permission by John A. Lindquist.

4. Clouds with Silver Linings: Europe Discovers the World
Edward Quin, A.D. 1498. The Discovery of America
    (1830)
Edward Quin, A.D. 1551. At the Death of Charles V
    (1830)
Image courtesy of the David Rumsey Map Collection.
www.davidrumsey.com

5. Ludacris and the "Ho"-Belt: A Rap Map of "Area Codes"
Stefanie Gray, Area Codes in Which Ludacris Claims
    to Have Hoes.
Image courtesy of Stefanie Gray.

6. The Antisaloon League Map of the World
The Anti-Saloon League Yearbook (1919), compiled
    and edited by Ernest Hurst Cherrington. Images
    courtesy of Joseph H. Eros.

7. Beautiful as a Snowflake: Fortress Coevorden
Plan of Fortress Coevorden.
Wikipedia map in the public domain.

8. National Dishes of Brazil
Photoillustration: Bruno Algarve and Ricardo Toscani.
Art Direction: Ken Tanaka.
Gula magazine.
"A Cara do Brasil."
Year: April 2006.
Ed: 162.

Photoillustration: Bruno Algarve and Ricardo Toscani
Art Direction: Ken Tanaka.
Sexy magazine.
"Bifão, ame-o ou deixe-o."
Year: September 2006.
Ed: 321.
http://www.flickr.com/photos/brunoalgarve/
www.toscani.art.br
http://www.flickr.com/photos/rtoscani/
http://www.flickr.com/photos/toscanhoto/

9. Great Britain as a Cloud
© Rob Gandy.

INDEX